Concepts in Geostatistics

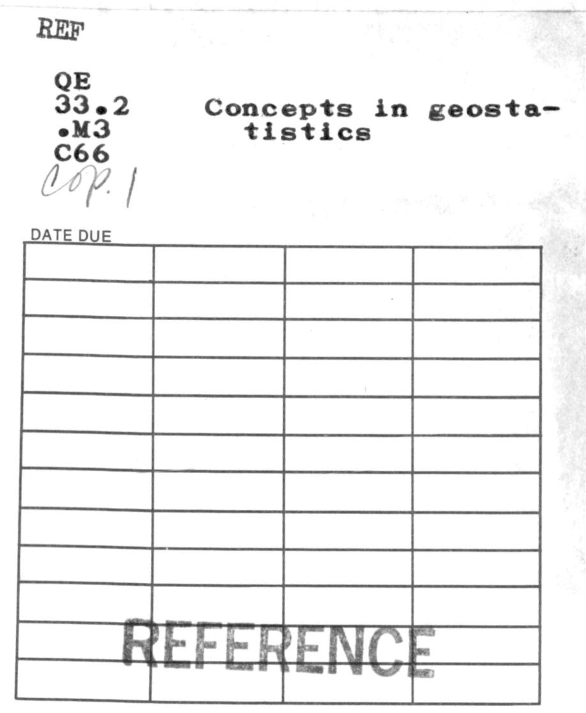

Concepts in Geostatistics

Edited by

Richard B. McCammon

Springer-Verlag
New York Heidelberg Berlin
1975

Library of Congress Cataloging in Publication Data

Main entry under title:

Concepts in geostatistics.

Includes bibliographies and index.
1. Geology--Statistical methods. 2. Electronic data processing--Geology. I. McCammon, Richard B., ed.
QE33.2.M3C66 519.5'02'455 74-23669

Printed in the United States of America

ISBN 0-387-06892-9 Springer-Verlag New York Heidelberg Berlin
ISBN 3-540-06892-9 Springer-Verlag Berlin Heidelberg New York

Preface

A two-week summer short course entitled Current Statistical Methods in Geology supported by the National Science Foundation was held at the University of Illinois at Chicago Circle in Chicago, Illinois from June 19 to June 30, 1972. The aim of the short course was to bridge the gap between the traditional first courses in statistics offered at most educational institutions and geostatistics as it is being developed by geologists and statisticians engaged in the application of statistics in geology. The course was intended for geology college teachers who were either then teaching or preparing to teach a course within their department dealing with computer applications and the use of statistical methods in geology.

This book arose out of the class notes which were prepared by the course director and the invited lecturers.

We are grateful to the 28 teachers who attended for their enthusiastic interest and thoughtful responses to the many statistical concepts presented to them as geologists during the two weeks of the course.

I am deeply grateful to my graduate assistants, Richard Kolb and Andrea Krivz, for the long hours spent in collating the course materials, testing the various computer programs, and instructing the participants in the use of computer BASIC.

Richard B. McCammon

Introduction

It is now little over 10 years since Miller and Kahns' Statistical Analysis in the Geological Sciences appeared in the geologic literature. By all accounts, this is considered to be the first modern text in statistical geology. Since then, Krumbein and Graybills' An Introduction to Statistical Models in Geology, the two volume work by Koch and Link, Statistical Analysis of Geological Data, and Davis' Statistics and Data Analysis in Geology have appeared. These books have been witness to the increasing quantification taking place in geology and the earth sciences generally. Coupled with the advances in computers, the geologist is now in a position to portray his data and characterize his results on a scale that heretofore was not possible. Briefly, the numeric treatment of geologic data has come of age.

In the quantification of geology, statistics has served as midwife to the concept of the process response model applied to geologic processes. Because precise hypotheses about natural processes have always proved difficult to formulate, it is not surprising that statistical rather than deterministic models have been put forward. Today statistics is being applied in virtually every branch of geology.

As elsewhere in science where statistics has been applied, however, what has held back the more rapid assimilation of statistical concepts in the minds of those engaged within a particular discipline is the absence of an orderly presentation of statistics as it applies to the particular discipline. Although this deficiency has been largely overcome in physics, chemistry, and lately, biology, this is not yet the case for geology. Moreover, there have come to be identified a number of statistical methods commonly used in geology that are not sufficiently understood by the average geology teacher so as to be presented effectively to the student.

There is little doubt that the geologist of the future will be required to make more quantitative judgments. In assessing the impact of technology on the environment for instance, the geologist will have to interpret data obtained from a wide variety of sources from which he will be expected to extract a more exact meaning. In reconstructing more precisely the Earth's past based on geophysical measurements and geochemical analyses of rock samples, he will need to perform statistical analyses in order that more exact inferences may be drawn from the data. Because more quantitative data will be collected, more quantitative geologic models will need to be developed. It is likely that as geologic prediction becomes more precise, it will become more quantitative. A geologist therefore will need to be better trained in the application of statistics.

The educational imperative for statistics in geology, therefore, is to introduce statistical concepts into the curriculum. This can be done either as a course in geostatistics, or, if this is not feasible, to incorporate basic statistical concepts into those geology courses that utilize statistical methods. While it is true that students in geology and related earth science fields must continue to be encouraged to take the basic course in statistics, the fact remains that exposure to statistics within the field of interest is the anvil upon which a more meaningful grasp of statistics will be forged.

Few geology departments today can afford the luxury of having one of its faculty members specialize in geostatistics. While it is recognized that a more quantitative approach to geology is evolving, there remains the more pressing problem of unifying the earth sciences and exposing the student to a more comprehensive view of the Earth's environment, past and present. Therefore, what is most likely to happen at this time is for a department to single out a member who has found statistics particularly useful in his field of study and to ask him to teach a course in geostatistics. Upon agreeing to this, the faculty member in question realizes soon afterward his own limited exposure to statistics or what is more likely, his inadequate knowledge of the application of statistics in geology in fields outside of his own. It was with this in mind that a two-week summer short course for geology college teachers was given and from which this book has evolved.

The book is divided into chapters corresponding to the material presented by the different lecturers in the course. There has been no attempt made to treat any subject in its full detail nor has there been a concerted effort to survey all the possible topics covering the field of statistical geology. The idea rather has been to introduce some basic concepts and to give examples of applications of statistics in geology with the intention of provoking interest and eventually generating discussion among geologists. For someone who is either now teaching or is planning to teach a course in geostatistics, it is hoped this book will serve as a guide. Much of the contained material was prepared specifically for the two week short course. For the student most of all, it is hoped that the book will make for enjoyable reading and fruitful study.

Richard B. McCammon

List of Contributors

Felix Chayes
Geophysical Laboratory
Carnegie Institute of Washington
Washington, DC 20008

William T. Fox
Department of Geology
Williams College
Williamstown, MA 01267

J.E. Klovan
Department of Geology
University of Calgary
Calgary 44, Alberta
Canada

W.C. Krumbein
Department of Geological Sciences
Northwestern University
Evanston, IL 60201

R.B. McCammon
Department of Geological Sciences
University of Illinois at Chicago Circle
Chicago, Illinois 60680

Daniel F. Merriam
Department of Geology
Syracuse University
Syracuse, NY 13210

Contents

Chapter 1

Statistics and Probability

R. B. McCammon

1.1 SAMPLE MEAN AND VARIANCE

Perhaps the best known statistic for describing a given set of numbers is the arithmetic mean. The mean conveys the notion of central tendency. With respect to data, the mean is linked with the idea of sampling and statistical inference. As familiar as the mean may be to most of us, its sequential properties may not be so well known. Thus, for ordered observations, whether these be based on time, space, or experimental design, the mean of n such observations is given by

$$\bar{x}_n = \frac{x_1 + x_2 + \ldots + x_n}{n}$$

where it is understood that the x_i-th observation precedes the x_j-th observation for $i<j$. If we add now an observation and recalculate the mean based on $(n + 1)$ observations, we can write

$$\bar{x}_{n+1} = \frac{x_1 + x_2 + \ldots + x_{n+1}}{n + 1} = \left(\frac{n}{n + 1}\right)\bar{x}_n + \left(\frac{1}{n + 1}\right)x_{n+1}$$

where x_{n+1} represents the new observation. This we recognize as the recursive form of the mean.

Next, we define ν

$$\nu = \frac{\bar{x}_{n+1}}{\bar{x}_n}$$

as the ratio of the mean calculated for n + 1 observations divided by the mean calculated for the preceding n observations, and similarly, we define

$$\eta = \frac{x_{n+1}}{\bar{x}_n}$$

where x_{n+1} represents the next observation. From this, we can write

$$\nu = \frac{n}{n+1} + \left(\frac{1}{n+1}\right)\eta$$

or, solving for η,

$$\eta = (\nu - 1)n + \nu$$

We can ask now how large or how small must a new observation be in order to affect the mean significantly. Suppose, for instance, we find that the mean is doubled after a new observation is added to ten previous observations. For this to happen, the new observation would have to be 12 times greater than the previous mean. In order to double a mean calculated from 100 preceding observations, the new observation would have to be 102 times greater than the preceding mean. We conclude, therefore, that the mean becomes increasingly more difficult to alter with increasing sample size unless there are increasingly erratic fluctuations in the observations. In the context of a time-dependent geologic process, we can conclude that cumulative effects tend toward equilibrium with advancing geologic time and that any significant departure from equilibrium is most likely due to outside influences.

Another statistic used to characterize a set of given values is the variance. The variance describes the scatter about the mean and for n observations, is defined as

$$s_n^2 = \frac{\sum_{i=1}^{n} (x_i - \bar{x}_n)^2}{n - 1}$$

where $\bar{x}_n$ represents the mean of the n observations. The denominator is given by n - 1 rather than n by virtue of the fact that the numerator can be expressed as the sum of n - 1 squared terms each of which is independent of the mean.

If we consider n-ordered observations as before, we can write

$$s_{n+1}^2 = \frac{n-1}{n} s_n^2 + \frac{1}{n+1} (x_{n+1} - \bar{x}_n)^2$$

the recursive form of the variance. If

$$x_{n+1} = \bar{x}_n$$

it follows that

$$s_{n+1}^2 < s_n^2 \qquad \text{for } s_n^2 > 0$$

This reinforces our earlier comment as regards the increase in the stability of the mean with an increase in sample size.

We turn the problem around now slightly and inquire for which new observation will it be true that

$$s_{n+1}^2 = s_n^2$$

Using the recursive relation, we obtain

$$x_{n+1} = \bar{x}_n \pm \sqrt{\frac{n+1}{n}}\, s_n$$

as the value for which the variance remains constant. It implies however that the new mean will be different. Thus the paradox is that in order to maintain a constant mean, the variance must be reduced, whereas to maintain a constant variance, the mean must change. For successive observations, the mean and variance cannot both remain constant (unless the variance is zero). For observations further removed, the mean and variance can both remain constant, however, if one considers cyclic fluctuations. This is, in fact, the definition of a stationary time series about which we shall hear more later.

From what we have said about the variance, it should not be difficult for you to write the recursive form for the covariance between two ordered pairs of observations. From there, it should not be much more difficult to write the recursive form of the correlation coefficient. This is left as an exercise.

1.2 ELEMENTS OF PROBABILITY

Probability can be viewed as partial information either known or presumed known prior to an event. In the previous section, the mean and variance were looked upon as descriptors of past data collected. In terms of probability, our concern lies with both past and future data. For some random variable Y, we associate the probability $p(y)$ that Y will take on the value y. Such a statement is conditioned by information H and thus we define the conditional probability $p(y|H)$ as the probability that Y takes on the value y given information H. An example from geology that illustrates the concept of conditional probability is the fossil collector in search of trilobites who estimates that there is a much greater probability of finding a trilobite in Cambrian strata compared with Cretaceous strata.

The two most important properties of conditional probability are that

$$p(y|H) \geq 0 \text{ for all } y \in Y$$

and

$$\int_{y \in Y} p(y|H)\, dy = 1$$

or

$$\sum_{y \in Y} p\,(y|H) = 1$$

if Y is discrete.

We can argue further that H represents information on a second random variable X; consequently, the conditional probability $p(y|x)$ is expressed as

$$p(y|x) = \frac{p(x,y)}{p(x)}$$

where $p(x,y)$ is the joint probability for X and Y and $p(x)$ is the unconditional probability for X. If Y is independent of X

$$p(x,y) = p(x)p(y)$$

so that

$$p(y|x) = p(y)$$

which is another way of saying that Y is independent of X.

For Y dependent on X, we can write

$$p(y) = \int_{x \in X} p(y|x)p(x)\,dx$$

for the unconditional probability of Y.

In many instances, it is necessary to transform one random variable to a new variable. For the continuous case, if

$$Y = f(X)$$

it follows that

$$p(y)\,dy = p(x)\,dx$$

for all $x \in X$ and $y \in Y$. Solving for Y, we have

$$p(y) = p(x)\,|dx/dy|$$

Let us consider an example. We define

$$p(x) = \begin{cases} 2x & 0 \le x \le 1 \\ 0 & \text{otherwise} \end{cases}$$

This is a probability density since

$$\int_0^1 p(x)\,dx = \int_0^1 2x\,dx = x^2 \Big|_{x=0}^{x=1} = 1$$

Suppose we wish to find the probability density of Y where

$$Y = X^2$$

Taking the derivative,

$$dy = 2x\,dx$$

we have

$$p(y) = (2y^{1/2}) \cdot |1/2y^{1/2}| = 1$$

so that

$$p(y) = \begin{cases} 1 & 0 \le y \le 1 \\ 0 & \text{otherwise} \end{cases}$$

Thus, the random variable Y is uniformly distributed between 0 and 1.

Another distribution we may need to establish is the sum of two independent random variables each of which follows the same probability density. In general, we can write for continuous variables

$$p(x + y) = \int_{u \in U} p(u)p(z - u)\, du$$

where Z = X + Y.

As an example, consider

$$p(x) = p(y) = \begin{cases} 1 & 0 \leq x,y \leq 1 \\ 0 & \text{otherwise} \end{cases}$$

We wish to find the probability density for Z defined as

$$Z = X + Y$$

We write

$$p(z) = \begin{cases} \int_0^z p(u)p(z - u)\, du & 0 \leq z \leq 1 \\ \int_{z-1}^1 p(u)p(z - u)\, du & 1 \leq z \leq 2 \end{cases}$$

or

$$p(z) = \begin{cases} z & 0 \leq z \leq 1 \\ 2 - z & 1 \leq z \leq 2 \end{cases}$$

as the probability density for Z. Try the following problems.

1.3 PROBLEMS

1.3.1 Consider the probability density function for X given by

$$f(x) = \begin{cases} 1 & 0 \leq x \leq 1 \\ 0 & \text{otherwise} \end{cases}$$

Let Y = $-\ell n$ X. Find the probability density function for Y.

1.3.2 Consider the probability density function for X given by

$$f(x) = (1/\sqrt{2\pi})e^{-1/2x^2} \qquad -\infty<x<+\infty$$

Let $Y = e^X$. Find f(y).

1.3.3 Let the probability density function for X and Y be given by

$$f(x) = \begin{cases} 1/a & 0\le x\le a \\ 0 & \text{otherwise} \end{cases}$$

$$f(y) = \begin{cases} 1/a & 0\le y\le a \\ 0 & \text{otherwise} \end{cases}$$

Let Z = X + Y. Find the probability density function for Z.

1.4 SEARCHING FOR DIKES

Suppose that we have a line segment of length L located somewhere inside a given area. We propose to locate the segment by conducting a search along parallel traverse lines spaced a distance D apart. We will consider the line segment found if one of the traverse lines intersects the segment. For $L \le D$, we ask then what will be the probability that the line segment is found. Within a geologic context, the line segment might represent the horizontal trace of a mineralized dike and the parallel line traverses represent the survey lines of a field party. Assuming the mineralization associated with the dike has economic value, the cost of such a search can be weighed against the expected value of the potential ore deposit. Ignoring the economic implications, consider the probabilistic aspect of the problem.

To say that a line segment of length L lies somewhere in a given area and nothing more presumes that such a line segment has a random orientation with respect to an arbitrarily chosen traverse line. For an arbitrary angle θ, the situation can be seen in Figure 1.1. Thus, the length component h of the line segment perpendicular to a given line of traverse is given by

$$h = L \sin\theta$$

and consequently, the probability of intersecting the line segment for fixed θ is given by

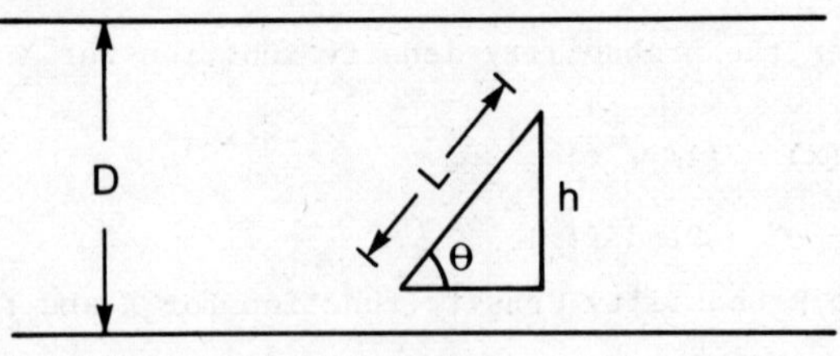

FIGURE 1.1

$$\Pr\{I|\theta\} = \frac{h}{D} = \frac{L}{D} \sin \theta$$

where $\Pr\{I|\theta\}$ represents the conditional probability of intersection. To obtain the unconditional probability $\Pr\{I\}$, we must integrate

$$\Pr\{I\} = \int_\theta \Pr\{I|\theta\}\Pr\{\theta\}$$

where $\Pr\{\theta\}$ defines the probability density for θ. The integration is performed for all values of θ. On the assumption that θ is randomly distributed, we define $\Pr\{\theta\}$

$$\Pr\{\theta\} = \frac{1}{\pi}d\theta \qquad 0\leq\theta\leq\pi$$

which is to say that θ is uniformly distributed between 0 and π.

By substitution, we obtain

$$\Pr\{I\} = \frac{2}{\pi}\frac{L}{D}\int_0^{\pi/2} \sin \theta \; d\theta = \frac{2}{\pi}\frac{L}{D}$$

as the probability of intersection.

In the search for the dike, we may wish to search on a grid. Thus, we can inquire as to the probability of intersecting the line segment of length L for a search conducted on a grid having mesh size D. For a fixed angle θ, the situation can be seen in Figure 1.2. Thus, we need to consider an intersection of the line segment with either horizontally or vertically spaced lines of the grid. If we take the horizontal to mean the x direction and the vertical to mean the y direction, then the probability of intersection $\Pr\{I\}$ is equal to

$$\Pr\{I\} = \Pr\{I_x \cup I_y\} = \Pr\{I_x\} + \Pr\{I_y\} - \Pr\{I_x \cap I_y\}$$

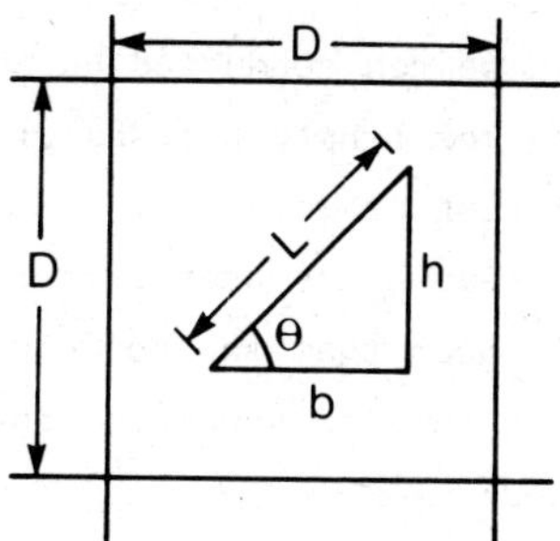

FIGURE 1.2

where $\{I_x \cup I_y\}$ represents the intersection along either a horizontal or vertical line of traverse and $\{I_x \cap I_y\}$ represents the intersection of both a horizontal and vertical line. Because the grid has a square outline, it follows that

$$\Pr\{I_x\} = \Pr\{I_y\}$$

and hence

$$\Pr\{I\} = \frac{4}{\pi}\frac{L}{D} - \Pr\{I_x \cap I_y\}$$

To derive the expression for the latter, we can write

$$\Pr\{I_x \cap I_y\} = \int_\theta \Pr\{I_x, I_y/\theta\}\Pr\{\theta\}$$

where $\Pr\{I_x, I_y/\theta\}$ is the conditional probability of intersecting the line segment with both a horizontal and vertical line of the grid. Referring to Figure 1.2,

$$\Pr\{I_x, I_y/\theta\} = \left(\frac{h}{D}\right)\left(\frac{b}{D}\right) = \left(\frac{L}{D}\right)^2 \sin\theta\cos\theta$$

so that

$$\Pr\{I_x \cap I_y\} = \frac{2}{\pi}\left(\frac{L}{D}\right)^2 \int_0^{\pi/2} \sin\theta\cos\theta\, d\theta = \frac{1}{\pi}\left(\frac{L}{D}\right)^2$$

Consequently, we have

$$\Pr\{I\} = \frac{4}{\pi}\frac{L}{D} - \frac{1}{\pi}\left(\frac{L}{D}\right)^2 = \frac{1}{\pi}\frac{L}{D}\left(4 - \frac{L}{D}\right)$$

as the probability that a search conducted on a grid with mesh size D will locate a line segment of length L ($L \leq D$) given that the line segment has no known orientation or location.

While the conducted search has been oversimplified in terms of actual practice, there is much that can be deduced from this simple example. After presenting this to students, for example, the following questions can be posed:

1. If, in fact, the line segment in question is suspected of having a known preferred orientation, how then can the probability of intersection based either on a parallel-line or grid-type search be maximized?
2. How is the probability affected if $L>D$? At this point it is prudent to pause and remind ourselves that some questions, though simply stated, cause considerable distress. The above question falls within this category. For $L>D$ in the case of parallel line search, for instance, the conditional probability of intersection $\Pr\{I/\theta\}$ is given by

$$\Pr\{I|\theta\} = \begin{cases} \frac{L}{D}\sin\theta & 0 \leq \theta \leq \sin^{-1} D/L,\ \sin^{-1}\left(-\frac{D}{L}\right) \leq \theta \leq \pi \\ 1 & \sin^{-1}\frac{D}{L} \leq \theta \leq \sin^{-1}\left(-\frac{D}{L}\right) \end{cases}$$

so that the unconditional probability of intersection $\Pr\{I\}$ is

$$\Pr\{I\} = \frac{2}{\pi}\int_0^{\sin^{-1}(D/L)} \frac{L}{D}\sin\theta\, d\theta + \int_{\sin^{-1}(D/L)}^{\pi/2} d\theta$$

$$= \frac{2}{\pi}\left[\frac{\pi}{2} - \sin^{-1}\frac{D}{L} + \frac{L}{D}\left(1 - \frac{\sqrt{L^2 - 1}}{L}\right)\right]$$

This is by way of saying that questions posed to the students must be thought out beforehand.

3. Does the probability of intersection change if, instead of a line segment, a circle of diameter L is considered?
4. Taking into account the economics of a search for a mineralized dike having an expected value V, for what spacing of D will the expected gain be maximized? In posing this question, it is

necessary to specify the cost of the search as a function of the line spacing or the grid size.

Thus far, our attention has focused on a single line segment or put into its geologic context, a single dike. In Figure 1.3, however, you will notice there are 100 such line segments (or dikes) of equal length that have been located at random within a square area. We can enlarge our original problem by asking for the probability of intersecting the i-th line segment with length L. This is equal to

$$\Pr\{I_i\} = \begin{cases} \frac{2}{\pi}\frac{L}{D} & \text{(for parallel lines spaced a distance D apart)} \\ \frac{1}{\pi}\frac{L}{D}\left(4 - \frac{L}{D}\right) & \text{(for a square grid with mesh size D)} \end{cases}$$

depending on the type of search. For example, consider the parallel-line type of search. If we assume that the location and the orientation of the different line segments with length L are each independent, it follows that the number of intersections observed for a given set

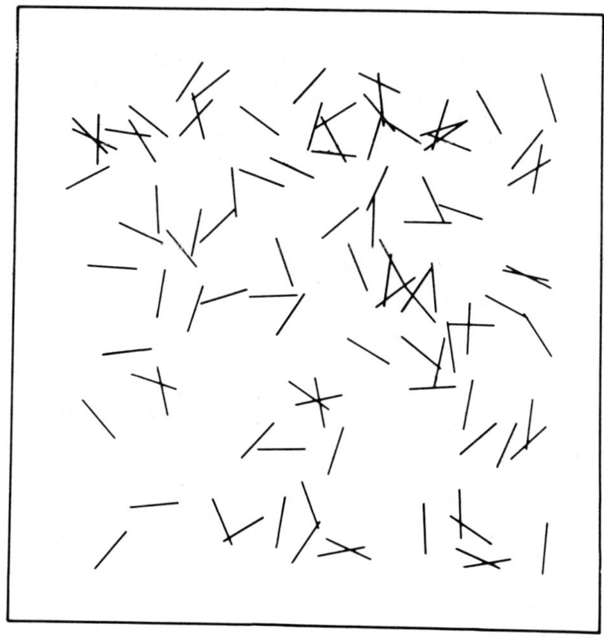

FIGURE 1.3

of parallel-line traverses spaced a distance D apart is binomially distributed. The expected number of such intersections N is given by

$$E(N) = NP\{I_i\}$$

where N is the total number of line segments contained within the search area.

Take a sheet of tracing paper and with a ruler make a series of parallel lines spaced a distance D, D = 2L, apart. Next, place this overlay on Figure 1.3 and for an arbitrarily chosen orientation, count the number of intersections of line segments with traverse lines. Repeat this several times if you wish, varying both the location and orientation of the overlay. Since the total number of line segments is known, you can compare the observed number with the expected number given in this instance as

$$E(N) = \frac{2}{\pi}\frac{NL}{D} = \frac{N}{\pi}$$

since L/D equals 1/2.

Under such conditions, therefore, when the total number of line segments of length L is unknown, the estimate of the total number can be based on the observed number of intersections given as

$$N \text{ est total} = \frac{D}{L}\frac{\pi}{2} N \text{ obs}$$

All we have said above applies equally to a grid type of search. A grid of mesh size D = 2L can be used to perform a similar experiment. Remember, the probability of an intersection is different than for parallel-line spacing however.

A question that can be posed to students at this point is what effect there is on the probability of an intersection if the length of the line segment or line segments to be located is unknown. It may be the case even that this length has a specified probability density function Pr{L}. Thus, the length L can be treated as a variable. The probability of an intersection P{I} is expressed in this instance by

$$\Pr\{I\} = \int_{\theta}\int_{L} \Pr\{I|L,\theta\}\Pr\{L,\theta\}$$

or if we assume that the length L is statistically independent of the angle θ,

$$\Pr\{L,\theta\} = \Pr\{L\}\Pr\{\theta\}$$

We can write

$$\Pr\{I\} = \int_L \int_\theta \Pr\{I|L,\theta\}\Pr\{L\}\Pr\{\theta\}$$

The probability of an intersection, therefore, is seen to vary depending on the distributions assigned to L and θ; consequently, L and θ become parameters that affect the observations made for different search strategies. A variety of probability density functions could be inserted in the above equation with the result that the probability of an intersection would differ from one situation to the next.

1.5 ROCKS IN THIN SECTION

We turn our attention now to a problem that takes us beyond the probability of an intersection. We consider the length of a line of intersection. Imagine that a circle of diameter D is located somewhere between two parallel lines spaced a distance t apart as shown in Figure 1.4. Suppose we locate another line at random that lies between the two lines and extends parallel to them. The probability that this line will intersect the circle is given by

$$\Pr\{I\} = D/t$$

where Pr{I} is the probability of intersection. Here we are interested not in the probability of intersection but rather in the length of

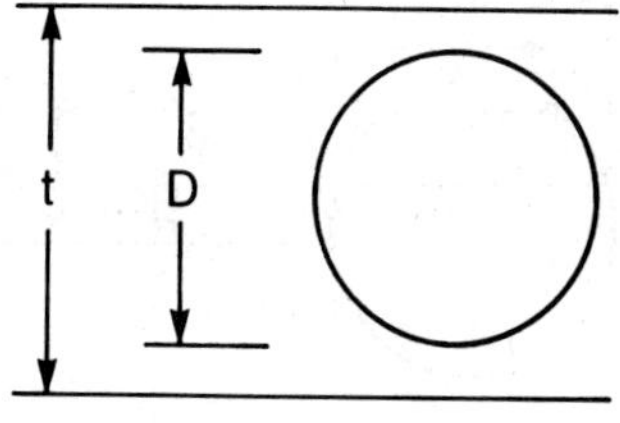

FIGURE 1.4

the chord of the circle being intersected. This chord can be considered the apparent diameter of the circle. We wish to derive the probability density of this length. To anticipate the geologic implication of this problem, it is sufficient to take note that the random slicing of circles by lines is identical in concept to the preparation of rock thin sections in which grains imbedded in a matrix are cut through by a random plane. For the latter, the grain size distribution observed subsequently in thin section will underestimate the actual particle size distribution in the rock. It is natural to examine what effect this has on the moments of the true distribution and how this bias, since it exists, can be reduced if not eliminated.

Taking the simplest case, we ask what is the probability, given that our circle is intersected by the line, that the observed length L will be greater than some length L_a ($L_a > 0$) and less than or equal to some length L_b ($L_a \leq L_b \leq D$). Referring to Figure 1.5, it is seen that this probability is given by

$$\Pr\{L_a < L \leq L_b | I\} = \frac{2h}{D}$$

where the multiplier of 2 derives from the presence of two slabs of thickness h occurring within the diameter of the circle. From the above figure, we see that

$$H_a^2 + \left(\frac{L_a}{2}\right)^2 = \left(\frac{D}{2}\right)^2$$

and

$$H_b^2 + \left(\frac{L_b}{2}\right)^2 = \left(\frac{D}{2}\right)^2$$

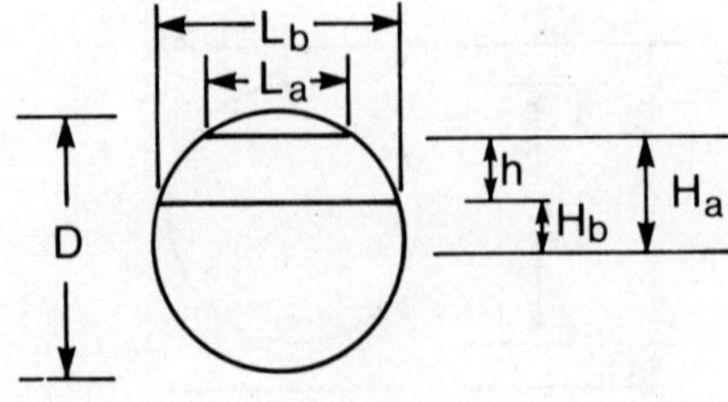

FIGURE 1.5

so that h is given by

$$h = H_a - H_b = \frac{D}{2}\left[\sqrt{1 - \frac{L_a}{D}^2} - \sqrt{1 - \frac{L_b}{D}^2}\right]$$

The unconditional probability that the observed length will lie within these limits is

$$\Pr\{L_a<L\le L_b\} = \Pr\{L_a<L\le L_b | I\}\Pr\{I\}$$

$$= \frac{D}{t}\left[\sqrt{1 - \left(\frac{L_a}{D}\right)^2} - \sqrt{1 - \left(\frac{L_b}{D}\right)^2}\right]$$

for

$$0\le L_a<L\le L_b\le D$$

If we let L_a approach 0 and set L_b to an arbitrary value $c\,(0<c\le D)$, the cumulative probability distribution for L is

$$\Pr\{L<c\} = \int_0^c p(\ell)\, d\ell = \frac{D}{t}\left[1 - \sqrt{1 - \left(\frac{c}{D}\right)^2}\right]$$

Taking the derivative with respect to L, the probability density is

$$p(\ell) = \frac{1}{t}\left[\frac{\ell/D}{\sqrt{1 - (\ell/D)^2}}\right] \quad \ell>0$$

As long as we are concerned with only a single diameter D, we can without loss of generality let t = D = 1 so that

$$p(\ell) = \frac{\ell}{\sqrt{1 - \ell^2}} \qquad \ell>0$$

for a circle of unit diameter. A graph of this probability density function is given in Figure 1.6. The distribution falls off rapidly for values much less than one. The question is, how much does this affect the estimate of the circle diameter if observations are based on the apparent diameters measured by successive random slices of a circle. The mean of the above distribution is given by

$$E(L) = \int_0^1 \ell p(\ell)\, d\ell = \int_0^1 \frac{\ell^2}{\sqrt{1 - \ell^2}}\, d\ell = \frac{\pi}{4}$$

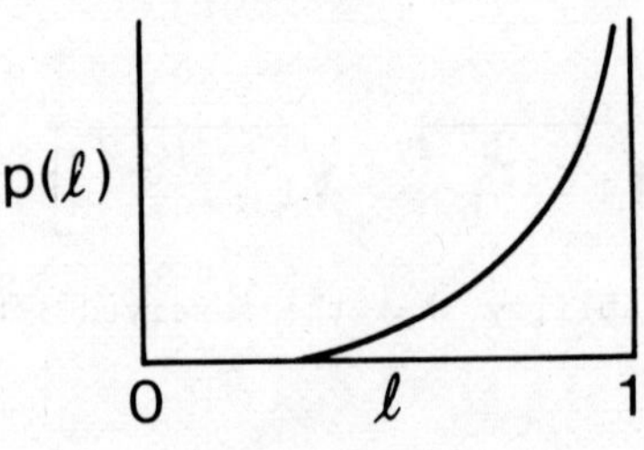

FIGURE 1.6

so that an estimate of the true diameter based on the average value of apparent diameters obtained by successive random slices would be in error by approximately 25 percent. In this instance, the estimate could be corrected simply by multiplying the average value of apparent diameter by $4/\pi$. In Figure 1.7, 50 circles of equal diameter D are located at random within a given square area. Using a ruled transparent overlay of parallel lines spaced a distance D apart measure the apparent diameters of circles intersected for several random positions of the overlay. The probability density of the apparent diameter of the circles is generated then. If you wish, construct a histogram for the values obtained. Next, calculate the mean and compare this with the expected value of the distribution above. The two values should agree within the precision allowed by the total number of measured intersections.

In reality, particles are of different sizes and mixed together; therefore we must consider a distribution of diameters. In the present context, this can be represented by circles having different diameters mixed together in fixed proportions.

To advance our discussion, we must take note that what we took before as $p(\ell)$, we now mean as $p(\ell|D)$, where D is a specified diameter. The unconditional probability density of the apparent diameter for mixtures of circles having different diameters can be expressed as

$$\Pr\{L\} = \int_D \Pr\{L|D\}\Pr\{D\}$$

where $\Pr\{D\}$ is the probability density of the circle having a diameter D. Rewritten in lower case, it is

$$p(\ell) = \int_d p(\ell|d)p(d)$$

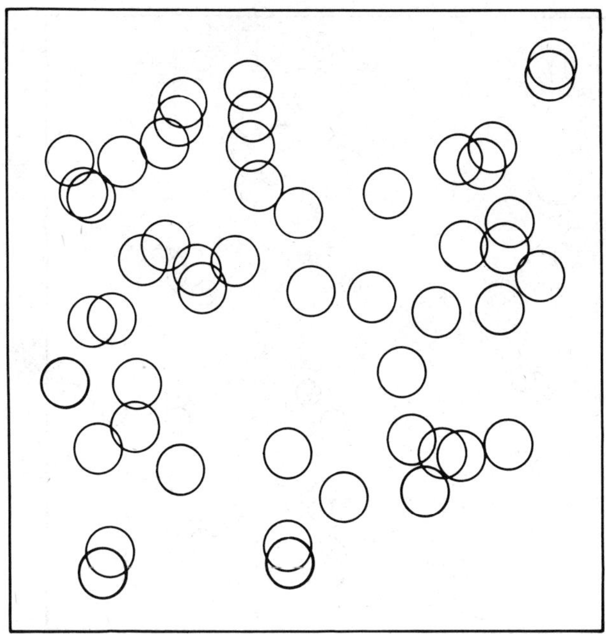

FIGURE 1.7

The general problem is now as follows: Given an observed distribution $p(\ell)$ of apparent diameters, find the distribution of the true diameters $p(d)$ given that $p(\ell|d)$ is specified. In Figure 1.8, for instance, 80 circles 1/4 inch in diameter and 20 circles 1/2 inch in diameter are located at random within a given square area*. Using an overlay with parallel lines spaced 1/2 inch apart, measure the apparent diameters of the circles intersected for several random orientations. Again, you may wish to construct a histogram. For this example, $p(d)$ is given by

$$p(d_i) = \begin{cases} 0.8 & d_1 = 0.25 \\ 0.2 & d_2 = 0.50 \\ 0 & \text{otherwise} \end{cases}$$

so that

$$p(\ell) = \frac{1}{t} \sum_{i=1}^{2} p(d_i) \frac{\ell/d_i}{\sqrt{1 - (\ell/d_i)^2}} \qquad \ell > 0$$

*Reduced scale used in Figure 1.8

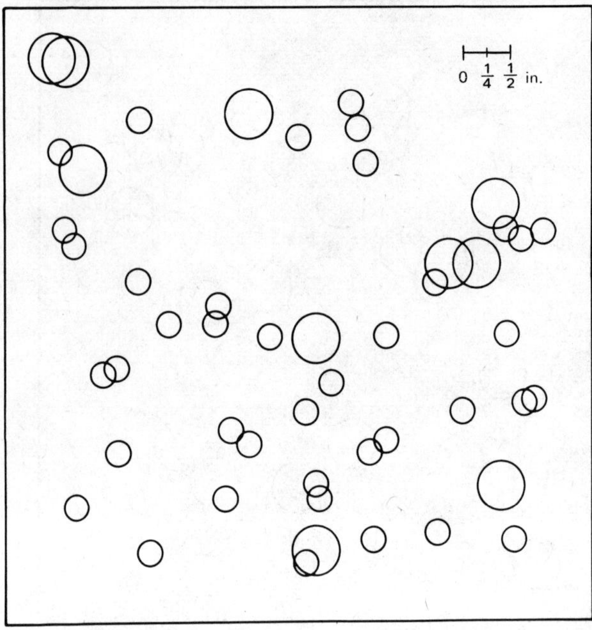

FIGURE 1.8

and therefore

$$E(L) = \frac{\int_{d_i} \ell p(\ell)\, d\ell}{\int_{d_i} p(\ell)\, d\ell} \qquad \ell>0$$

$$= \frac{\frac{1}{d_2} \sum_{i=1}^{2} \frac{p(d_i)}{d_i} \int_0^{d_i} \frac{\ell^2\, d\ell}{\sqrt{1-(\ell/d_i)^2}}}{\frac{1}{d_2} \sum_{i=1}^{2} p(d_i) d_i}$$

$$= \frac{(\pi/4)\,[p(d_1)\frac{d_1^2}{d_2} + p(d_2)d_2]}{p(d_1)\frac{d_1}{d_2} + p(d_2)}$$

$$= \frac{(\pi/4)[0.8 \times (1/8) + 0.2 \times (1/2)]}{0.8 \times (1/2) + 0.2}$$

$$= \frac{1}{3}\frac{\pi}{4}$$

where t has been set equal to the largest diameter. Compare the mean of the apparent diameters with this expected value. Once again, the values should agree within the precision allowed by the total number of measured intersections. While the solution to the general problem where there is a continuous distribution of particle diameters is obviously more difficult, the same principles apply as to this example.

These examples, as with the examples in the preceding section, have touched lightly on the more broader topic of geometric probability. Readers who wish to pursue details of the subject further can refer to the short but lucid monograph by Kendall and Moran (1963) and two more recent review articles by Moran (1966, 1969). Within the field of geology, geometric probability has been applied to the study of particle size in thin section (Rose, 1968) and to the probability of success of locating elliptical targets underground with square, rectangular, and hexagonal grids (Singer, 1972). For these references, particularly the latter two, the reader will recognize the difficulty in translating the relatively straightforward equations discussed in this chapter to the more complex situations met with in practice. Despite these difficulties, however, the concept of geometric probability offers a very practical device for studying the uncertainty associated with spatial form in geology.

REFERENCES

Kendall, M. G., and Moran, P. A. P., 1963, Geometrical probability: London, Chas. Griffin, Std., 125 p.

Moran, P. A. P., 1966, A note on recent research in geometrical probability: Jour. App. Prob., v.3, p. 453-563.

Moran, P. A. P., 1969, A second note on recent research in geometrical probability: Adv. App. Prob., v. 1, p. 73-90.

Rose, H. E., 1968, The determination of the grain size distribution of a spherical granular material embedded in a matrix: Sediment., v. 10, p. 293-309.

Singer, G. A., 1972, Ellipgrid, A FORTRAN IV program for calculating the probability of success in locating elliptical targets with square rectangular and hexagonal grids: Geocom. Programs, v. 4, p. 1-10.

Chapter 2

R- and Q-Mode Factor Analysis

J. E. Klovan

2.1 MEANING OF FACTOR

Factor analysis is a generic term that describes a variety of mathematical procedures applicable to the analysis of data matrices. Although developed, and largely exploited by psychologists, it is a method of general application to many branches of scientific enquiry -- and geology is no exception.

At the outset the word "factor" requires precise definition because the way it is interpreted can give a false impression as to what the method attempts to do. Mathematically, a factor refers to one of a number of things that when multiplied together yield a product. Another use of the word is in reference to some sort of theoretical or hypothetical casual variable. As will become clear, it is the former meaning that should be applied to the method; occasionally the second meaning may be applicable to the results of the method.

The principles of the mathematics involved in factor analysis were outlined by Pearson in 1901. Starting in 1904, Spearman began applying the method to psychological theories. Thurstone, Holzinger, and a large number of other workers expanded on the method during the 1930's and 1940's. The advent of electronic computers in the 1950's made the laborious calculations involved amenable to quick solution and the methods became widely available. In the late 1950's the method was first applied to geologic problems.

Geologists are commonly faced with problems wherein a large number of properties are measured or described on a large number of things. The "things" may, for example, be rocks and the "properties" may be the amounts of various minerals making up the rocks. If these data are arranged in tabular form such that each rock represents a row of the table and each mineral species a column, then the resulting chart of numbers is referred to as a data matrix.

Analysis of such a data matrix may pose a considerable problem to the investigator if it contains many numbers. The primary aim of factor analysis is to achieve a parsimonious description of such a data matrix -- that is, to determine if this table of numbers can be simplified in some way.

Returning to rocks and minerals as a concrete example, perhaps there are a small number of mineral assemblages, which, if determined, describe the rocks almost as well as all the amounts of the individual minerals. In this case the objective of factor analysis would be to simplify the original large data matrix by determining

1. The number of mineral assemblages present
2. The composition of each assemblage in terms of the original mineral species
3. A description of each rock sample in terms of the amount of each assemblage present in it

The present chapter will attempt to outline the mathematical procedures used in one of the methods of factor analysis, namely, the method of principal components. A simplified heuristic approach will be followed that will attempt to make use of the geologists' ability to visualize three-dimensional concepts. Several simple examples will be used to lead the reader through a formidable mathematical jungle, and finally, some real applications of the method are briefly explained.

For readers with no experience with matrix algebra, the Appendix contains concepts that may be helpful in the following exposition.

2.2 DATA MATRICES

A matrix is a table of numbers with so many rows and so many columns. As a matter of convention here, the rows of a data matrix will represent geologic entities, the columns will represent attributes of these entities. In most cases there will be more entities than attributes so that most data matrices are rectangular in shape and are "taller" than they are "wide." More simply, data matrices tend to have more rows than columns.

In the terminology of matrix algebra, an entire matrix is symbolized by a capital letter. "X" will be used to symbolize any data matrix. The size of the matrix is specified by a double subscript notation, thus $X_{N,n}$ refers to a table of numbers with N rows and n columns. If 93 rocks have been analyzed for 12 minerals, the resulting data matrix may be symbolized as $X_{93,12}$.

The entities of a geologic data matrix will depend on the nature of the problem. Rock or sediment specimens are obvious cases. Samples of water or oil collected from various formations are also common. (Note that the word "sample" raises some semantic problems in that it carries a special statistical connotation.)

Attributes, often referred to as variables, also depend on the nature of the problem. A rock may be analyzed as to its mineral components in which case the amount of each mineral is considered an attribute. The rock could equally well (or, as well) be analyzed in terms of certain chemical elements. The amount of an element then becomes an attribute.

Clearly, attributes do not exist in and of themselves; they are properties of things. It is important, therefore, to define at the outset of an investigation what is an entity and what is an attribute. A fossil, for example, may in one study be considered an entity and various features of it will be attributes. Or, in another study, the amount of that fossil in a stratum may be considered an attribute of that stratum.

2.3 FACTOR ANALYTIC MODES

Confronted with a data matrix the investigator may focus his attention on two distinct yet interrelated questions:

R-mode: If the primary purpose of the investigation is to understand the inter-relationships among the attributes, then the analysis is said to be an R-mode problem.

Q-mode: If the primary purpose is to determine interrelationships among the entities, then the analysis is referred to as Q-mode.

In many cases both R- and Q-mode analyses are performed on the same data matrix. As discussed later, factor analysis is applicable to both types of questions. The essential solutions of the factor analysis are only slightly dependent on the mode. The exact nature of this relationship is described in more complete detail in a recent paper (Klovan and Imbrie, 1971).

2.4 THE R-MODE MODEL

Given the data matrix $X_{N,n}$, the basic problem is to determine m linear combinations of the original n variables that describe the geological entities without significant loss of information (assuming m<<n). These m linear combinations are termed factors. The method of analysis operates not on the original data matrix but rather on the matrix of correlation coefficients derived from the data matrix.

The well-known Pearson product-moment correlation coefficient is the standard means of assessing the degree of linear relationship between a pair of variables.

If X_i and X_j are any two variables, that is, two columns from the data matrix X, then the correlation coefficient between them may be computed from:

$$r_{ij} = \frac{\sum_{k=1}^{N} (X_{ki} - \overline{X}_i)(X_{kj} - \overline{X}_j)}{\sqrt{\sum_{k=1}^{N} (X_{ki} - \overline{X}_i)^2}\sqrt{\sum_{k=1}^{N} (X_{kj} - \overline{X}_j)^2}} \qquad (2.4.1)$$

where the notation $\sum_{k=1}^{N}$ refers to summation over all the entities; $\overline{X}_i$ and $\overline{X}_j$ are the mean values of variable X_i and X_j.

This is the so-called raw score formula and the situation is portrayed in Figure 2.1.

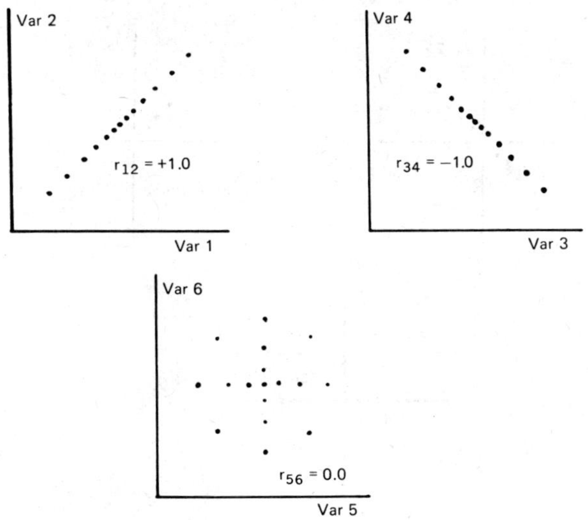

FIGURE 2.1

The origin of this graph may be shifted without changing the configuration of the points. If the mean value of a variable is subtracted from every value of the variable, the results are <u>deviate scores</u>. The resulting numbers show how far from the mean each entity is. This also results in shifting the origin of the variable to its mean value.

If we define x_i and x_j as two variables in deviate form, then the correlation formula becomes

$$r_{ij} = \frac{\sum_{k=1}^{N} x_{ki} x_{kj}}{\sqrt{\sum_{k=1}^{N} x_{ki}^2} \sqrt{\sum_{k=1}^{N} x_{kj}^2}} \tag{2.4.2}$$

This situation is shown on Figure 2.2

A variable in <u>standard form</u> is defined as

$$Z_i = \frac{X_i - \overline{X}_i}{\sigma_i} = \frac{x_i}{\sigma_i} \tag{2.4.3}$$

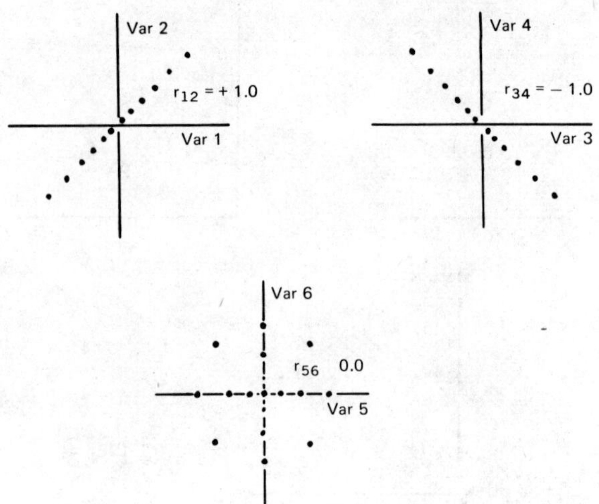

FIGURE 2.2

where σ_i is the standard deviation of variable X_i. A standardized variable may be viewed as having a mean of zero and a standard deviation of one. The individual values of the variable show how far from the mean an entity is in terms of units of standard deviation.

The standard deviation of a variable is given by

$$\sigma_i = \sqrt{\sum_{k=1}^{N} \frac{(X_{ki} - \overline{X})^2}{N}} = \sqrt{\sum_{k=1}^{N} \frac{x_{ki}^2}{N}} \tag{2.4.4}$$

or

$$\sqrt{N}\sigma_i = \sqrt{\sum_{k=1}^{N} x_{ki}^2} \tag{2.4.5}$$

(Editor's note: This definition differs slightly from the one used in Chapter 1 in that N, rather than N-1, is in the denominator. In factor analysis, the sample size is usually large enough so that this difference can be safely ignored.)

Thus

$$Z_i = \frac{X_i - \overline{X}_i}{\sigma_i} = \frac{x_i}{\sigma_i} = \frac{x_i}{\sqrt{\frac{1}{N}} \sqrt{\sum_{k=1}^{N} x_{ki}^2}} \tag{2.4.6}$$

Substituting this into formula (2.4.2) we obtain

$$r_{ij} = \frac{1}{N}\sum_{k=1}^{N} z_{ki}z_{kj} \tag{2.4.7}$$

This formula illustrates the fact that the correlation coefficient is nothing more than the average value of the cross-product between two variables given in standard form.

Up to now only two variables have been considered. In the general case, correlation coefficients are computed between every possible pair of variables and arranged into a square symmetrical matrix $R_{n,n}$. This matrix contains all the information regarding the pairwise linear relationships between the variables.

In Figure 2.1 the correlation coefficient was perceived to measure the degree of linear association between two variables as measured by the scatter of data points. Note that the axes of the graph are the variables and the entities are points on the graph. If three variables are considered, then the third variable is constructed as an axis at right angles to the other two, and the entities will form some sort of three-dimensional swarm of points. Other variables can be added by constructing axes at right angles to all other axes but of course this situation cannot be portrayed in three dimensions. A _row_ of the data matrix may then be considered as a vector that gives the coordinates of an entity in n-dimensional space.

The situation may be reversed. A graph may be constructed using the entities as sets of orthogonal axes as in Figure 2.3. Here the variables become points on the graph. A _column_ of the data matrix may then be considered as a vector that gives the coordinates of a variable in N-dimensional space.

If the variables are expressed in deviate form, that is the origin is at the mean, then variable x_i and variable x_j can be portrayed as two vectors in N-space. From the Pythagorean theorem, the length of a vector is equal to

$$\ell_i = \sqrt{\sum_{k=1}^{N} x_{ki}^2} \tag{2.4.8}$$

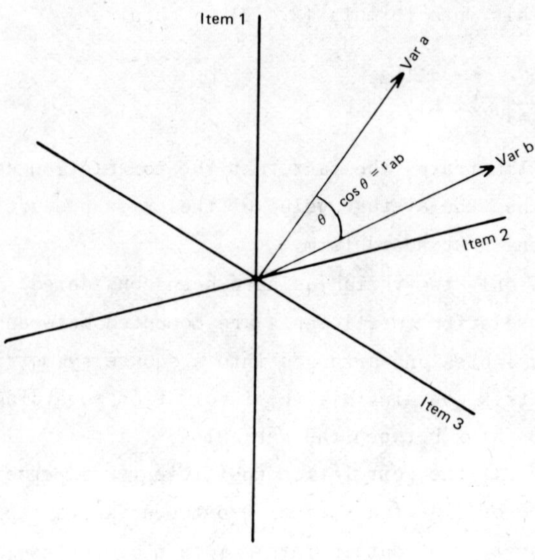

FIGURE 2.3

Further, elementary trigonometry shows that the angle θ between the two vectors is equal to

$$\cos\theta = \sum_{K=1}^{N} \frac{x_{ki}x_{kj}}{\ell_i \ell_j} = \frac{\sum_{k=1}^{N} x_{ki}x_{kj}}{\sqrt{\sum_{k=1}^{N} x_{ki}^2}\sqrt{\sum_{k=1}^{N} x_{kj}^2}} \qquad (2.4.9)$$

which is exactly equivalent to formula (2.4.2).

Thus the correlation coefficient between any two variables is also the cosine of the angle between the two vectors representing the variables situated in N-space.

Both interpretations of the correlation coefficient will be found useful in the following discussion.

The following equation perhaps best summarizes the underlying rationale of factor analysis:

$$Z_j = a_{1j}F_1 + a_{2j}F_2 + \cdots + a_{mj}F_m + a_jE_j \qquad (2.4.10)^*$$

In words, the equation states that any variable (for convenience considered in standard form) Z_j, consists of a linear combination of m common factors plus a unique factor. The resemblance of this equation to a multiple regression equation should be obvious.

In the factor model, the F's refer to hypothetical variables called factors. It is assumed that each of these m factors will be involved in the delineation of two or more variables, thus the factors are said to be common to several variables; m is assumed to be less than n, the number of variables. The a's are analogous to β weights in regression analysis. They are weights to be applied to the factors so that the factors can best predict the value of Z_j; "best" defined in a least-squares sense. In factor analysis parlance, the a's are termed loadings and the F's factor scores. The factor designated E_j is a factor unique to variable Z_j and is analogous to the error term in a regression equation.

The factor model contains n such equations; one for each variable. For a particular entity k, equation (2.4.10) becomes:

$$Z_{kj} = a_{1j}F_{k1} + a_{2j}F_{k2} + \cdots + a_{mj}F_{km} + a_jE_{kj} \qquad (2.4.11)$$

The values for the a's do not change from entity to entity (just as the β's remain constant in regression equations), but the values of the F's do change from entity to entity. An excellent way to view the F's is to think of them as new variables that are linear combinations of the old variables. As such, each entity can "contain" a different amount of each one of these new variables. The F's are referred to as factor scores.

The basic problem then is threefold:

1. To determine values for the a's
2. To determine values for the F's
3. To determine m, the number of common factors

*In this equation and the following derivation, the author realizes that he is confusing the principal component model with that of a true factor analytic model. The justification for this is that, in practice, most geologic applications follow this model, and, additionally, it is easier to explain the underlying rationale and objectives in this form.

There are several ways in which an explanation of the solution can be approached. Two such approaches will be dealt with here.

The equation for the variance of a variable in standard form is given by

$$\sigma_j^2 = \frac{\sum_{k=1}^{N} Z_{kj}^2}{N} \tag{2.4.12}$$

Due to the standardization process the variance of Z_j is, of course, equal to one.

In terms of the factor model the variance may be written as:

$$\sigma_j^2 = \frac{\sum_{k=1}^{N} Z_{kj}^2}{N} = \frac{a_{1j}^2 \Sigma F_{k1}^2}{N} + \frac{a_{2j}^2 \Sigma F_{k2}^2}{N} + \cdots + \frac{a_{mj}^2 \Sigma F_{km}^2}{N} + \frac{a_j^2 \Sigma E_{kj}^2}{N}$$

$$+ 2\left(\frac{a_{1j}a_{2j}\Sigma F_{k1}F_{k2}}{N} + \frac{a_{1j}a_{3j}\Sigma F_{k1}F_{k3}}{N} \ldots\right) + \frac{a_{mj}a_j \Sigma F_{km}E_{kj}}{N} =$$

$$= 1 \tag{2.4.13}$$

Two simplifying restrictions may now be imposed.

1. The factors must be in standard form.
2. The factors must be uncorrelated.

The first constraint makes every term of the form $\Sigma F_{k1}^2/N$ equal to one (since this is the variance of the factor).

The second constraint makes every term of the form $\Sigma F_{k1}F_{kp}/N$ equal to zero [see equation (2.4.7)].

The entire equation thus becomes:

$$\sigma_j^2 = a_{1j}^2 + a_{2j}^2 + \cdots + a_{mj}^2 + a_j^2 \tag{2.4.14}$$

It is seen that the total variance of a variable is to be made up of the sum of the squared a's.

Further, the total variance consists of two parts.

1. That due to the common factors. This is termed the <u>communality</u>, symbolized h_j^2.

$$h_j^2 = a_{1j}^2 + a_{2j}^2 + \cdots + a_{jm}^2 \tag{2.4.15}$$

2. That due to the unique factor. This of course is equal to 1 - h_j^2 and by definition is that part of the variance of variable j that is not shared by any of the other variables. It is analogous to the error term in regression analysis.

The method of principal components attempts to minimize this unique variance in the solution for the a's and F's.

The algebraic notation of the factor model is very cumbersome and not readily comprehended. Matrix notation allows an easier representation of the model.

As has been pointed out, the $X_{N,n}$ data matrix can be transformed to the standardized version $Z_{N,n}$.

We can consider the Z matrix as being the sum of two matrices:

$$Z_{N,n} = C_{N,n} + E_{N,n} \tag{2.4.16}$$

where C contains the "true" measures and the matrix E contains "error" measures.

It is a fact that any matrix can be expressed as the product of two other matrices. Thus the matrix Z can be considered as the product of the matrix F and A or

$$Z = FA' \tag{2.4.17}$$

where A' is the transpose of A.

The F matrix contains the factor scores; the A matrix the factor loadings. But the model contains two types of F's and A's, the common and unique portions.

The factor loading matrix can be considered as consisting of two parts.

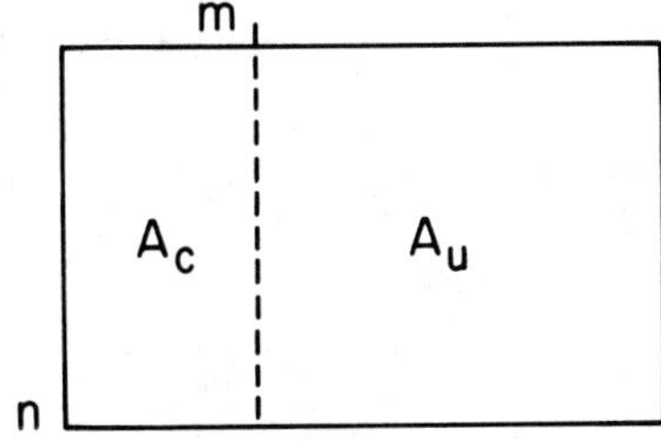

The first m columns contain the common factor loadings; the last u columns contain the unique factor loadings; A_u is a diagonal matrix.

Similarly, the F matrix can be partitioned into a part containing m columns of common factor scores and a part containing u columns of unique factor scores.

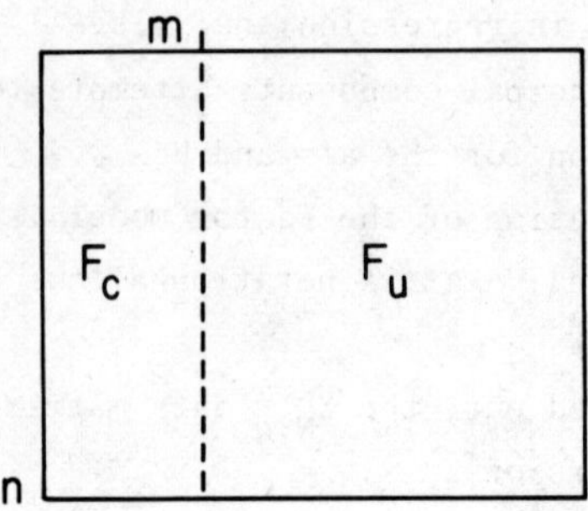

The model is now evident; A_c and F_c will be chosen in such a way as to yield the matrix C; A_u and F_u yield the matrix E. The sum of C and E, of course, yield the original matrix Z.

To summarize, the total data matrix is considered to be derivable from the product of two other matrices ($Z = FA'$).

Z can further be considered as the sum of two matrices C and E; C containing "true" measures, E containing error measures: $Z = C + E$. Both F and A can be partitioned into two components, a common variance part and an error part. Thus

$$Z = F_c A'_c + F_u A'_u \tag{2.4.18}$$

Because we are usually only interested in the matrix C, solution for F_c and A_c will be sufficient. E can always be obtained from $E = Z - C$.

The basic matrix manipulations required for solution are presented below in point form. Each step is then explained with reference to a simple geometric model.

1. The correlation matrix may be obtained from

$$R = \frac{1}{N} Z'Z \tag{2.4.19}$$

2. The basic factor model states that

$$Z = FA' \tag{2.4.20}$$

or

$$Z' = AF' \tag{2.4.21}$$

3. Substituting (2.4.20) and (2.4.21) into (2.4.19) and ignoring the constant $\frac{1}{N}$, we obtain

$$R = Z'Z = AF'FA' \tag{2.4.22}$$

4. We impose the condition that the factors will be uncorrelated; that is, F will be orthonormal.

$$F'F = I \tag{2.4.23}$$

where I is the identity matrix. Thus, (2.4.22) becomes

$$R = AA' \tag{2.4.24}$$

5. We impose the constraint that

$$A'A = \Lambda \tag{2.4.25}$$

where Λ is the diagonal matrix of eigenvalues of the correlation matrix R, or

$$U'RU = \Lambda \tag{2.4.26}$$

where U contains the eigenvectors associated with Λ. U is a square orthonormal matrix so that:

$$U'U = UU' = I \tag{2.4.27}$$

6. The following matrix manipulation provides the solution.

(a) Pre-multiply (2.4.26) by U

$$UU'RU = U\Lambda$$

$$RU = U\Lambda \tag{2.4.28}$$

(b) Post-multiply (2.4.28) by U'

$$RUU' = U\Lambda U'$$

$$R = U\Lambda U' \tag{2.4.29}$$

(c) Because R is a square symmetric matrix with the Gramian property (positive semi-definiteness):

$$R = U\Lambda U' = U\Lambda^{1/2}\Lambda^{1/2}U' \tag{2.4.30}$$

7. Substituting into (2.4.25)

$$R = AA' = U\Lambda^{1/2}\Lambda^{1/2}U'$$

$$A = U\Lambda^{1/2} \tag{2.4.31}$$

8. The matrix F may be solved for from

$$Z = FA'$$
$$ZA = FA'A$$
$$ZA = F\Lambda$$
$$F = ZA\Lambda^{-1} \tag{2.4.32}$$

Explanation

<u>Step 1</u>. Earlier it was shown that this equation was valid. Geometrically, Figure 2.4 shows the scatter diagram interpretation of a situation involving three variables. Note that the swarm of data points is in the form of a three-dimensional ellipsoid. Theoretically, every correlation matrix will define such an ellipsoid -- a hyperellipsoid when more than three dimensions are involved.

<u>Step 2</u>. The basic factor equation states that the data matrix can be considered as the product of two matrices F and A. Unfortunately, matrix theory shows that there is an infinite number of pairs of matrices F and A whose product will reproduce Z.

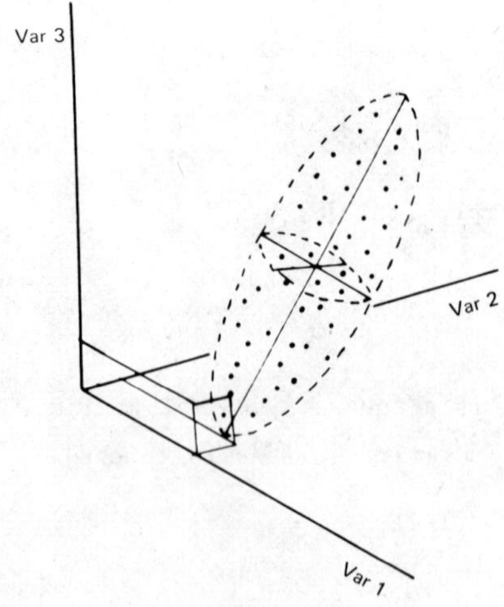

FIGURE 2.4

Step 3. This equation simply shows the relation between F, A, R, and Z.

Step 4. Because there is an infinite number of pairs of matrices whose product will yield Z, we impose a constraint that the F matrix be orthonormal. Simply, this means that factors will be in standard form and furthermore, they will be uncorrelated. If we consider these factors to be new variables then this implies that the new variables have no mutual correlation among them. Because of this constraint, it is seen that the F matrix can for the moment be disregarded from further consideration.

Steps 5-6. The crux of the method of principal components is embodied in equation (2.4.25).

Any square symmetric matrix, such as R, can be uniquely defined in terms of two other matrices that have special properties.

In the expression $R = U\Lambda U'$, Λ is a diagonal matrix containing the eigenvalues of R. U is a square orthonormal matrix containing the associated eigenvectors. The calculation of eigenvalues and eigenvectors is a straightforward matter using computer programs. Essentially, eigenvalues are the roots of a series of partial derivative equations set up so as to maximize the variance and retain orthogonality of the factors. Physically, the eigenvectors merely represent the positions of the axes of the ellipsoid (or hyperellipsoid) shown on Figure 2.4. The eigenvalues are proportional to the lengths of these axes. The largest eigenvalue and its corresponding eigenvectcr represent the major axis of the ellipsoid. It is important to note that the data points show maximum spread along this axis, that is, the variance of the data points is at a maximum.

The second largest eigenvalue and its eigenvector represent the largest minor axis. The axis is, of course, at right angles to the major acis and the data points are seen to have the second largest amount of variance along this direction. The same reasoning applies to the remaining eigenvalues and eigenvectors.

So what is accomplished at this step is to create a new frame of reference for the data points. Rather than using the old set of variables as reference axes we can use the eigenvectors instead. These have the property that they are located along directions of maximum variance and are uncorrelated.

Step 7. The equation R = AA' suggests that the correlation matrix derived from the original data can be duplicated exactly by the major product moment of the factor loadings matrix. This is true if as many factors as original variables are used. However, the matrix A_c of equation (2.4.18) will approximate R. The difference matrix, $R - A_c A'_c$ contains the residual correlations not accounted for by the common factors. The determination of the number of common factors needed is left for later discussion.

The end result of the matrix manipulations is the equation $A = U\Lambda^{1/2}$. This simply means that the desired matrix of factor loadings is the orthogonal matrix of eigenvectors of R, each column of which is scaled by the square root of the corresponding eigenvalue. This is merely a normalization process.

Step 8. The matrix of factor scores is obtained by straightforward matrix manipulation.

2.5 A PRACTICAL EXAMPLE

To recapitulate what has been discussed in rather abstract terms, and to give physical significance to the method as explained thus far, a simple geologic problem will be followed through.

Figure 2.5 is a typical geologic data matrix with 20 rows and 10 columns. The rows represent 20 localities and the columns, as indicated, represent attributes of the rocks and structures at each locality. The data are fictitious.

Figure 2.6 is the 10 x 10 correlation matrix obtained from these data.

Figure 2.7 is the list of eigenvalues obtained from the correlation matrix.

Geological Properties

Locality	Mg in calcite	Fe in sphalerite	Na in muscovite	Sulfide	Crystal size of carbonates	Spacing of cleavage	Elongation of ooliths	Tightness of folds	Veins/meter^2	Fractures/meter^2
1	1175	999	975	625	158	262	437	324	431	433
2	936	820	813	575	267	379	478	413	411	428
3	765	711	716	599	457	548	579	558	491	513
4	624	598	600	542	471	515	531	520	490	500
5	417	422	422	432	444	441	437	439	437	437
6	401	403	375	401	405	270	317	290	515	465
7	520	504	488	469	427	370	410	386	507	482
8	661	626	618	553	462	466	506	480	529	523
9	877	787	773	594	354	401	493	434	500	498
10	1060	932	898	656	315	312	468	370	580	552
11	1090	960	935	681	334	375	518	427	567	555
12	896	811	790	629	403	411	511	448	570	555
13	748	688	672	560	401	399	472	426	525	512
14	617	573	553	477	360	315	385	342	487	462
15	436	424	389	393	361	207	277	236	514	455
16	664	587	560	419	212	182	287	221	397	369
17	750	665	651	484	259	299	387	331	399	396
18	903	797	791	573	291	396	486	427	421	437
19	998	888	887	657	366	499	583	527	480	506
20	1162	999	994	671	252	404	539	450	449	471

Data matrix

FIGURE 2.5

Correlation coefficients between the ten geological properties

Correlation Coefficients

Variable	1	2	3	4	5	6	7	8	9	10
1	1.000	0.998	0.994	0.908	-0.576	0.130	0.581	0.282	0.012	0.258
2	0.998	1.000	0.998	0.933	-0.523	0.183	0.625	0.334	0.057	0.313
3	0.994	0.998	1.000	0.942	-0.497	0.235	0.664	0.383	0.035	0.312
4	0.908	0.933	0.942	1.000	-0.180	0.477	0.834	0.610	0.286	0.590
5	-0.576	-0.523	-0.497	-0.180	1.000	0.616	0.258	0.519	0.539	0.550
6	0.130	0.183	0.235	0.477	0.616	1.000	0.880	0.987	0.181	0.524
7	0.581	0.625	0.664	0.834	0.258	0.880	1.000	0.944	0.216	0.604
8	0.282	0.334	0.383	0.610	0.519	0.987	0.944	1.000	0.208	0.573
9	0.012	0.057	0.035	0.286	0.539	0.181	0.216	0.208	1.000	0.909
10	0.258	0.313	0.312	0.590	0.550	0.524	0.604	0.573	0.909	1.000

Figure 2.6

Eigenvalues of correlation matrix

	Eigenvalue	Percent Variance Explained
Factor I	5.46	54.61
Factor II	3.19	86.54
Factor III	1.35	100.00

FIGURE 2.7

There are only three nonzero eigenvalues (that there are three and only three is a reflection of the "cooked-up" nature of the problem). Geometrically, the analysis began with the 20 data points scattered in 10-dimensional space. Because there are only three nonzero eigenvalues, the implication is that the hyperellipsoid enclosing the data points has seven axes of zero length and exists, in fact, as an ordinary three-dimensional ellipsoid. Therefore the data points can be located with reference to three mutually perpendicular axes instead of the original 10.

The A matrix, in Figure 2.8, will be found to reproduce R exactly from $R = AA'$. The F matrix in Figure 2.9 will, in conjunction with A, reproduce the standardized version of the data matrix according to $Z = FA'$.

The geologic interpretation of these matrices will be deferred until some additional concepts are put forward.

Principal Component Factor Matrix

		Factors		
	Comm.	1	2	3
1	1.0000	0.8029	-0.5894	0.0886
2	1.0000	0.8385	-0.5367	0.0940
3	1.0000	0.8579	-0.5122	0.0407
4	1.0000	0.9760	-0.1961	0.0943
5	1.000	0.0176	0.9998	-0.0098
6	1.0000	0.6538	0.5999	-0.4611
7	1.0000	0.9297	0.2393	-0.2799
8	1.0000	0.7647	0.5018	-0.4042
9	1.0000	0.3268	0.5407	0.7751
10	1.0000	0.6641	0.5437	0.5132
Variance		54.614	31.928	13.459
Cum. var		54.614	86.542	100.000

Principal Factors of Correlation Matrix

FIGURE 2.8

Principal Factor Score Matrix

Locality	Factors 1	2	3
1	0.3887	-2.2383	0.1359
2	0.1989	-0.9798	-1.1363
3	0.9083	1.2182	-1.0941
4	0.2926	1.3964	-0.9847
5	-0.9595	1.0959	-1.4845
6	-1.6185	0.6760	0.9116
7	-0.7464	0.9150	0.1871
8	0.2992	1.2964	0.0020
9	0.4987	0.0377	0.1130
10	0.9111	-0.4040	2.1310
11	1.2932	-0.1946	1.5206
12	0.8987	0.6101	1.1914
13	0.1935	0.5946	0.4475
14	-0.8159	0.1323	0.3074
15	-1.8532	0.1726	1.3474
16	-1.7613	-1.5741	-0.2310
17	-0.8554	-1.0525	-0.9248
18	0.2300	-0.6998	-1.1155
19	1.3189	0.1578	-0.7823
20	1.1785	-1.1592	-0.5417

FIGURE 2.9

2.6 FACTORS

The basic factor equation, Z = FA' states that the data matrix can be considered as the product of two factors F and A. The mathematical procedure used to obtain these matrices has been outlined, but it is now necessary to explain what they signify and how they are interpreted and used.

The matrix of factor scores, F, in general, consists of N rows and m columns where N is the number of entities and m equals the number of common factors. Each column is in standard form with zero mean and unit variance, and there is zero correlation between columns.

Because the factors are linear combinations of the original variables, they can themselves be considered as new variables with the abovementioned properties. Scanning down a column of F, the "amount" of this new variable as contained in each entity is revealed. Being in standard form, factor scores are expressed in units of standard deviation from the mean of the hypothetical variable. Thus, the variation from entity to entity is expressed in relative terms only. Nonetheless, these new variables can be plotted and manipulated in the same way as any other variable. Using the factors as orthogonal axes, the entities may be plotted on a scattergram to show the distribution of entities in m-dimensional space.

The matrix of factor loadings, A, generally has n rows and m columns. The rows correspond to the original variables; the columns are the factors. Each column has been scaled so that the sum of squared elements in the column is equivalent to the amount of original variance accounted for by that factor. The elements in a column may be considered as the coefficients of a linear equation relating the variables to the factor -- in essence, they give the recipe for the factor. Therefore, the columns of the A matrix can be used to give some physical meaning to the factors.

A row of the A matrix shows how the variance of a variable is distributed among the factors. Interrelationships between variables can be determined by a comparison of their rows in the A matrix.

As was pointed out in equation (2.4.15), the sum of the factor loadings squared in a row of A is an expression of the amount of variance of a variable accounted for by the m factors. This was termed the <u>communality</u>. The communality attached to each row of the A matrix gives an appreciation of how well each variable is explained by the m factors considered.

Another valid view of an element in A is that it represents the correlation between a variable and a factor. Because correlations are angular measures, the row elements actually represent the cosines between a variable and the m reference factor axes. A useful way to analyze and interpret the factors is to plot the loadings as two-dimensional scattergrams. For m factors there will be m(m - 1)/2 such graphs, which in effect gives two-dimensional snapshots of m-space. Groupings of variables and trends between them often yield important clues as to the physical significance of the factors.

A practical example of interpretation is deferred until the matter of rotation has been discussed.

It has been stressed several times that there is an infinite number of solutions for the equation Z = FA'. This is so because there is an infinite number of pairs of matrices F and A that will reproduce Z.

The method of principal components determines a unique solution because certain constraints are imposed, namely, that the F matrix is orthonormal and that the A matrix contains the eigenvectors of the correlation matrix produced from Z. These constraints yield a solution with two desirable properties; the factor axes are orthogonal and pass through positions of maximum variance. But, they also possess an undesirable property. Considered as new variables, they are very general and, in fact, correspond to a sort of average of all the original variables. Although this may in some instances be useful, it is common to move the positions of the factor axes by rotating them so that they will satisfy certain other criteria. An attempt is made to achieve what is termed "simple structure," by which is meant that the factor axes are located in positions such that:

1. For each factor only a relatively few variables will have high loadings, and the remainder will have small loadings.
2. Each variable will have loadings on only a few of the factors.
3. For any given pair of factors, a number of variables will have small loadings on both factors.
4. For any given pair of factors, some of the variables will have high loadings on the second factor but not on the first.
5. For any given pair of factors, very few of the variables will have high loadings on both.

What these conditions attempt to achieve is to place the factor axes in more meaningful positions, that is, so that they will be highly correlated with some of the original variables.

A large number of methods have been designed to accomplish these objectives but only two will be considered here.

An approximation to simple structure, designed by Kaiser (1958), uses a rigid rotation procedure. This means that the orthogonal principal component factors will be rigidly rotated and maintained orthogonal.

Kaiser's approach is to find a new set of positions for the principal factors such that the variance of the factor loadings on each factor is a maximum; the loadings should tend toward unity and zero (the sum of the variance for all m factors is the actual quantity maximized). That is, when the value of V in the following expression is maximized simple structure should be obtained:

$$V = n\sum_{p=1}^{m} \sum_{j=1}^{n} \left(\frac{b_{jp}}{h_j}\right)^4 - \sum_{p=1}^{m} \left(\sum_{j=1}^{n} \frac{b_{jp}^2}{h_j^2}\right)^2 \qquad (2.6.1)$$

where b_{jp} is the loading of variable j on factor p on the new, rotated factor axes.

The full explanation of this equation is rather too involved for these notes, and the reader is referred to Harman (1960, p. 301) for a full discussion. The process can be readily understood in terms of matrix algebra. Given the n by m matrix of principal factor loadings A, the objective is to transform it to an n by m matrix of varimax factor loadings B such that B will satisfy equation (2.6.1).

Usually, factors are transformed (or rotated) two at a time. In matrix terms, this can be accomplished by

$$B = AT \qquad (2.6.2)$$

where T consists of

$$\begin{bmatrix} \cos\phi & -\sin\phi \\ \sin\phi & \cos\phi \end{bmatrix}$$

ϕ is the angle of rotation required to yield a maximum value of V in equation (2.6.1) and is determined by an iterative process.

The matrix B contains the loadings of the original n variables on the m rotated factors and can be interpreted in the same way as the A matrix.

Factor scores for the varimax factors can also be computed. Again the values can be interpreted in the same way as the principal factor scores. The varimax factor scores remain in standard form but the factors are now slightly correlated.

Figure 2.10 shows the rotated, varimax factor matrix and Figure 2.11 the associated factor scores.

Figure 2.12 illustrates and compares the plots of principal components and varimax loading derived from the data matrix of Figure 2.5.

Varimax Factor Matrix

Var	Comm.	Factors 1	2	3
1	1.0000	0.9971	0.0765	-0.0060
2	1.0000	0.9916	0.1241	0.0362
3	1.0000	0.9835	0.1804	0.0117
4	1.0000	0.8813	0.3985	0.2540
5	1.0000	-0.6197	0.5880	0.5198
6	1.0000	0.0558	0.9897	0.1317
7	1.0000	0.5191	0.8380	0.1680
8	1.0000	0.2102	0.9648	0.1580
9	1.0000	0.0146	0.0488	0.9987
10	1.0000	0.2338	0.3979	0.8872
Variance		44.771	33.318	21.912
Cum. var		44.771	78.089	100.000

FIGURE 2.10

Note that in both cases, graphs of the factor loadings reveal similar patterns but with the varimax loadings the factor axes are located near the extreme clusters of variables. Interpretation is thus facilitated.

As can be seen from the graph and scanning down the columns of the varimax factor loading matrix B, loadings on variables 1, 2, and 3 are extremely high on Factor 1 and very low on the other two factors. This would lead to the interpretation that the "new" variable, Factor 1, is in some way a paleotemperature index because variables 1, 2, and 3 are all geologic paleothermometers. The first column of the varimax factor score matrix thus shows the distribution of paleotemperature at each of the 20 localities in terms of units of standard deviation away from the mean paleotemperature. Similarly, Factor 2, of the B matrix, may be interpreted in terms of deformation, while Factor 3 represents some index of permeability.

Varimax Factor Score Matrix

Locality	Factors		
1	1.7191	-1.1345	-0.9480
2	0.6130	0.2141	-1.3682
3	-0.2291	1.8610	0.0226
4	-0.7962	1.5403	0.0196
5	-1.6301	0.9332	-0.8890
6	-1.5345	-1.0766	0.6182
7	-1.1157	-0.0212	0.4175
8	-0.5908	0.9143	0.7663
9	0.3711	0.2439	0.2588
10	1.2434	-0.9398	1.7595
11	1.3197	-0.2480	1.4876
12	0.4639	0.1764	1.5324
13	-0.1684	0.1884	0.7211
14	-0.6634	-0.5747	0.0782
15	-1.3364	-1.7591	0.6394
16	-0.3829	-1.7847	-1.5171
17	-0.1144	-0.5587	-1.5440
18	0.4618	0.3838	-1.1944
19	0.7982	1.3086	-0.1605
20	1.5620	0.3336	-0.6997

FIGURE 2.11

This rather simple example illustrates the main features of R-mode factor analysis. Several complications arise when real data are analyzed, and these will be touched on following a discussion of Q-mode techniques.

2.7 THE Q-MODE MODEL

When the nature of a geologic problem is such that relationships between entities are the focus of attention rather than relationships

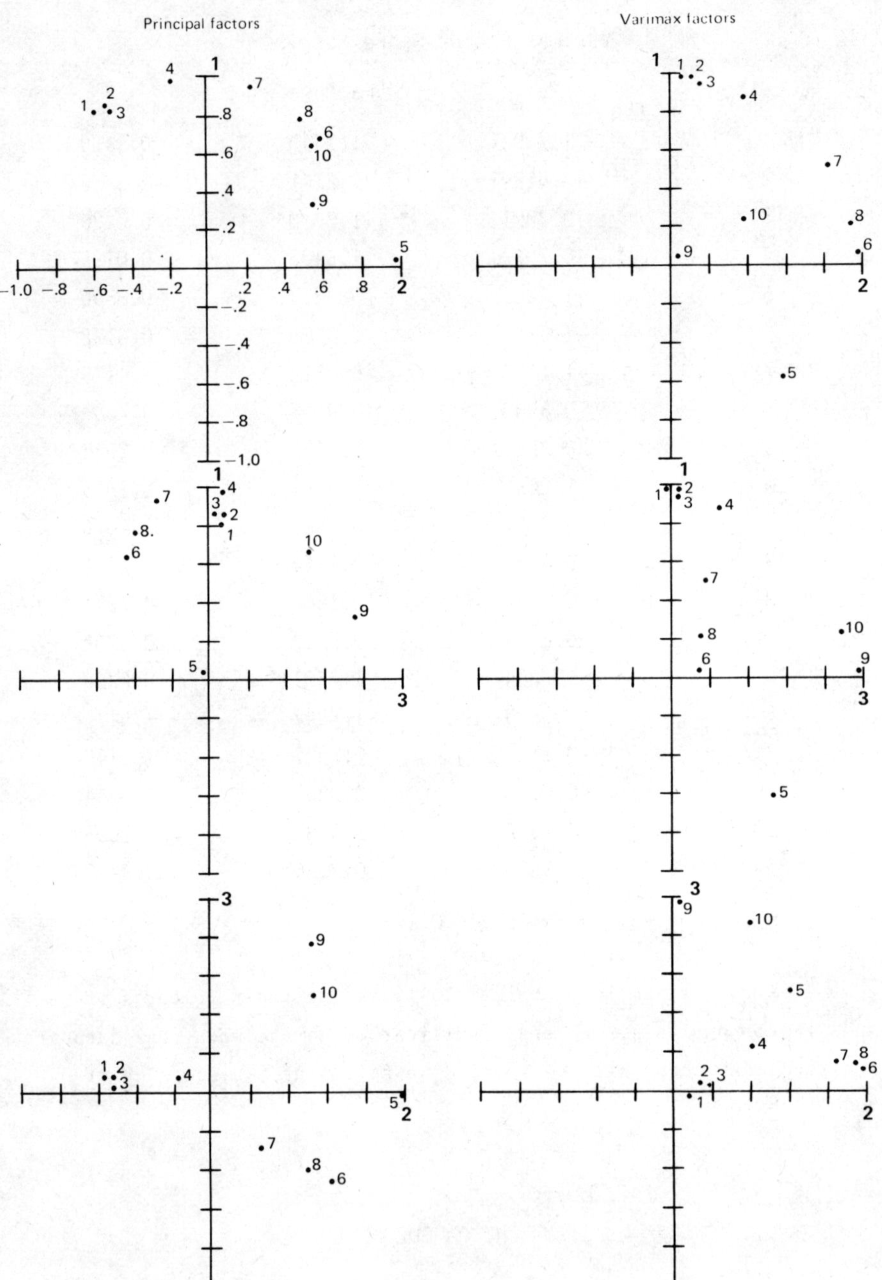
Principal factors
Varimax factors
1
2
3
.8
.6
.4
.2
−1.0 −.8 −.6 −.4 −.2
.2 .4 .6 .8
−.2
−.4
−.6
−.8
−1.0

FIGURE 2.12

between properties, then the Q-mode method of factor analysis becomes a useful analytical tool.

Numerous such geologic situations are easily envisaged. The delineation of lithofacies or biofacies are perhaps the most evident. Here the objective is to find groups of entities that are similar to one another in terms of their total composition.

The catch in this objective is to define "similarity" in a mathematically realistic way. Several measures of similarity will be discussed next. Once all the interentity similarities are computed and arranged in matrix form, the previously described methods of solution are applicable to the analysis of this similarity matrix.

Similarity Indices

There is a vast, and ever expanding literature, on the problems associated with a mathematical definition of similarity. There is little in the way of theoretical justification for the selection of one index over another; however, Gower (1967) has at least underscored several considerations that must be taken into account.

Aside from similarity coefficients designed for presence-absence data, three indices have commonly been used in Q-mode analysis.

1. Correlation Coefficient. The Pearson product moment correlation coefficient has been used to indicate the degree of relationship between two entities. If X_k and X_ℓ are any two rows of the data matrix then:

$$r_{k\ell} = \frac{\sum_{i=1}^{n} (X_{ki} - \overline{X}_k)(X_{\ell i} - \overline{X}_\ell)}{\sqrt{\sum_{i=1}^{n} (X_{ki} - \overline{X}_k)^2} \sqrt{\sum_{i=1}^{n} (X_{\ell i} - \overline{X}_\ell)^2}} \qquad (2.7.1)$$

measures the degree of relationship between the two entities.

As appealing as this index may appear there are a number of intuitive and theoretical drawbacks. Note, for example, the $\overline{X}_k$ and $\overline{X}_\ell$ terms. These are average values for item k and ℓ. If an item is described in terms of a wide variety of properties measured in different scales, what then does such an average mean in physical terms? By subtracting this value from each of the attribute quantities, the proportions of the attributes are altered from entity to entity. To partially overcome this, some workers advocate standardizing the data matrix by columns before using this equation. Clearly, this only adds a further complication to the data and complicates interpretation.

For this and several other reasons, the correlation coefficient is not considered a good index of similarity.

2. Coefficient of Proportional Similarity. Imbrie and Purdy (1962) define an index of similarity referred to as cos θ. The equation used is

$$\cos \theta_{k\ell} = \frac{\sum_{i=1}^{n} X_{ki} X_{\ell i}}{\sqrt{\sum_{i=1}^{n} X_{ki}^2} \sqrt{\sum_{i=1}^{n} X_{\ell i}^2}} \qquad (2.7.2)$$

For positive data this index ranges from zero, for perfect dissimilarity to one for complete similarity. The difficulty with this index is that while it preserves the proportional relationships between entities it is blind to the absolute sizes involved. Thus a "midget" and a "giant" whose attribute proportions are identical would be considered as being completely similar. In many problems where the investigator is interested in changes in the proportions of constituents, such as sedimentologic and faunal studies, the index is very appropriate. Imbrie (1963, p. 26) and McIntrye (1969, p. DBM-A-41) suggest methods for including a "size" variable that helps to remove the inherent "size" blindness of the index.

3. Distance Coefficient. Harbaugh (1964) describes the use of a coefficient that measures the distance between entities in n-dimensional space. The complement of this distance is then taken as a measure of similarity. In order to standardize this index, all attributes must be scaled so that the maximum value of each is 1 and the minimum value is zero. This, of course, distorts

the proportionality. The equation for computation, assuming scaled attributes, is:

$$d_{k\ell} = 1 - \sqrt{\frac{\sum_{i=1}^{n} (X_{ki} - X_{\ell i})^2}{N}} \tag{2.7.3}$$

However obtained, the similarities between all possible pairs of entities are calculated and arranged in a square, symmetric, similarity matrix $S_{N,N}$. This matrix contains all the information concerning the interrelations between the N entities under study. Q-mode factor analysis begins at this point.

It will be recalled that in R-mode analysis, the objective was to create m linear combinations of the n variables. The m linear combinations could be considered as new variables. Similarly, in Q-mode analysis, the objective is to find new, hypothetical entities whose compositions are linear combinations of those of the original entities.

As pointed out by Imbrie (1963), these "new" entities, or Q-mode factors, can be conceived of as being composite end members, combinations of which can be used to reconstruct the original entities. The problem then is to "unmix" the original entities into the smallest possible number of end members. In this respect, Q-mode analysis is a "mirror image" of R-mode analysis.

Computational Procedure

Using the cos θ measure of similarity, the following equations reveal the necessary steps in the analysis.

Let $X_{N,n}$ be the data matrix. Form the diagonal matrix D whose principal diagonal contains the square root of the row vector lengths of X. That is,

$$d_{kk} = \sqrt{\sum_{i=1}^{n} X_{ki}^2} \tag{2.7.4}$$

The matrix operation

$$W = D^{-1}X \tag{2.7.5}$$

row-normalizes each row of X, that is, each row vector in W is of unit length.

Then the similarity matrix is computed from

$$S = WW' \tag{2.7.6}$$

The basic factor equation is

$$W = AF' \tag{2.7.7}$$

and

$$W' = FA' \tag{2.7.8}$$

thus,

$$S = AF'FA' \tag{2.7.9}$$

The constraint

$$F'F = I \tag{2.7.10}$$

results in

$$S = AA' \tag{2.7.11}$$

As in the R-mode, we stipulate that

$$A'A = \Lambda \tag{2.7.12}$$

and following the same reasoning as before

$$A = U\Lambda^{1/2} \tag{2.7.13}$$

and

$$F = W'A\Lambda^{-1} \tag{2.7.14}$$

It must be emphasized that the A and F matrices so derived are <u>not</u> the same as those derived in the R-mode analysis. One set is derived from S, the other from R.

Once obtained, A may be rotated to B as discussed in the section on varimax rotation.

The B matrix in general consists of N rows and m columns. Each row corresponds to an entity; each column represents a factor.

The factors are best thought of as hypothetical entities that are completely dissimilar in terms of the proportions of their constituents.

Scanning down any column of B shows the amount of the hypothetical entity contained in each real entity. Scanning across any row shows the composition of a real entity in terms of the hypothetical entities.

The F matrix has n rows and m columns. The rows represent the original attributes used to describe the entities. The numbers in a row thus describe the relative amount of the attribute in each factor. A column gives the "composition" of the hypothetical entity in terms of the original attributes. Unfortunately the scale of these numbers is obscure. Thus, they can be used in relative terms only.

2.8 AN EXAMPLE

Imbrie (1963) presents an example of Q-mode analysis that gives the basic ideas behind the method. The matrix of fictitious data is given in Figure 2.13. The rows represent 10 sediment samples and the columns represent 10 species of minerals. The cos θ similarity matrix is given in Figure 2.14. Eigenvalues derived from this matrix indicate that there are three, and only three, independent dimensions to the data.

Data Matrix for Q-mode Example

	Var A	Var B	Var C	Var D	Var E	Var F	Var G	Var H	Var I	Var J
Loc 1	5.0	25.0	15.0	5.0	5.0	20.0	10.0	5.0	5.0	5.0
Loc 2	10.0	30.0	17.0	17.0	8.0	8.0	5.0	4.0	1.0	0.0
Loc 3	3.0	6.0	10.0	13.0	25.0	15.0	13.0	8.0	5.0	2.0
Loc 4	7.5	27.5	16.0	11.0	6.5	14.0	7.5	4.5	3.0	2.5
Loc 5	4.6	21.2	14.0	6.6	9.0	19.0	10.6	5.6	5.0	4.4
Loc 6	3.8	13.6	12.0	9.8	17.0	17.0	11.8	6.8	5.0	3.2
Loc 7	8.3	26.6	15.9	14.2	9.1	11.1	6.8	4.6	2.2	1.2
Loc 8	6.1	22.7	14.6	10.2	9.9	15.4	9.1	5.3	3.8	2.9
Loc 9	7.6	24.2	15.2	13.8	10.8	11.8	7.6	5.0	2.6	1.4
Loc 10	3.9	10.3	11.2	12.6	21.3	14.8	11.9	7.3	4.6	2.1

FIGURE 2.13

Cos theta matrix

	Loc 1	Loc 2	Loc 3	Loc 4	Loc 5	Loc 6	Loc 7	Loc 8	Loc 9	Loc 10
Loc 1	1.0000									
Loc 2	0.8739	1.0000								
Loc 3	0.6906	0.6480	1.0000							
Loc 4	0.9654	0.9704	0.6906	1.0000						
Loc 5	0.9888	0.8733	0.7908	0.9595	1.0000					
Loc 6	0.8849	0.8020	0.9479	0.8697	0.9445	1.0000				
Loc 7	0.9244	0.9901	0.7227	0.9902	0.9315	0.8723	1.0000			
Loc 8	0.9714	0.9391	0.7941	0.9862	0.9860	0.9390	0.9779	1.0000		
Loc 9	0.9282	0.9779	0.7758	0.9855	0.9457	0.9082	0.9967	0.9867	1.0000	
Loc 10	0.7878	0.7531	0.9877	0.7952	0.8705	0.9828	0.8200	0.8790	0.8636	1.0000

FIGURE 2.14

The varimax factor loading matrix and its associated factor score matrix are given in Figures 2.15 and 2.16.

Figure 2.17 is one of three possible plots of the varimax factor loading. It is now evident that the original 10 sediment samples are various mixtures of the three hypothetical samples. Three maps could be drawn showing the spatial distribution of the end members and from this, mechanism of transport might be deduced.

The composition of the end-members can be roughly determined from a study of the columns of the F matrix of Figure 2.16.

Varimax Factor Matrix

	Commun	Factors 1	2	3
Loc 2	1.0000	0.9133	0.3492	0.2095
Loc 7	1.0000	0.8494	0.4311	0.3043
Loc 4	1.0000	0.8179	0.3810	0.4311
Loc 9	1.0000	0.8080	0.5006	0.3107
Loc 8	1.0000	0.7266	0.5183	0.4510
Loc 1	1.0000	0.6605	0.3899	0.6416
Loc 5	1.0000	0.6228	0.5222	0.5826
Loc 3	1.0000	0.3094	0.9314	0.1918
Loc 10	1.0000	0.4363	0.8632	0.2541
Loc 6	1.0000	0.4900	0.7714	0.4060
Variance		47.56	36.05	16.40
Cum percent		47.56	83.60	100.00

FIGURE 2.15

Varimax Factor Score Matrix

	Factors 1	2	3
Var A	0.881	0.038	-0.294
Var B	2.446	-0.468	0.948
Var C	1.135	0.422	0.483
Var D	1.333	1.003	-1.346
Var E	-0.097	2.433	-0.742
Var F	-0.206	0.971	2.167
Var G	-0.177	1.061	0.811
Var H	0.030	0.668	0.199
Var I	-0.208	0.393	0.611
Var J	-0.218	0.086	0.809

FIGURE 2.16

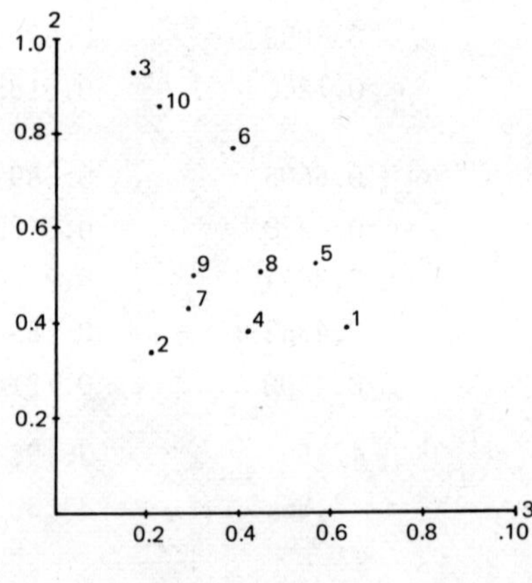

FIGURE 2.17

2.9 OBLIQUE ROTATION

The factors obtained in R- and Q-mode analysis are constrained to be orthogonal. There may be no physical reason for them to be mutually orthogonal and thus many schemes have been devised to find sets of factors that are oblique to one another.

Of these, the method due to Imbrie (1963) is the most simple. Referring to Figure 2.17, it is apparent that the most divergent real sediment samples 1, 2, 3 could be used as end members from which all the remaining samples could be derived. Imbrie's method is to rotate the factor axes so that they coincide with the most divergent samples and then express all the other samples as proportions of these end member samples.

This is accomplished by constructing an m by m matrix T, which contains the varimax loadings of the most divergent samples. Then

$$C = BT^{-1} \tag{2.9.1}$$

yields the oblique factor matric C. Figure 2.18 illustrates the results of the operation applied to the problem just discussed. The method is also applicable to R-mode factor loadings matrices. The oblique factors are no longer uncorrelated and computation of factor scores becomes much more involved.

In both of the examples presented thus far the correct choice of the number of factors needed to reproduce the data matrix has been unequivocable. Coincidentally, in both instances there were three, and only three, nonzero eigenvalues. But both of these examples involved fictitious data that were "manufactured" for the purpose of exposition. When analyzing real data the investigator must select the correct number of factors, and this is seldom unequivocable. Although some statistical methods are available to aid in this section, experience has shown that certain empirical criteria are more useful.

The essentials of the problem are:

1. In order to reproduce exactly the data matrix from the two derived matrices (Z = FA') it is necessary to use as many factors as original variables. This is because the common variance and unique variance together equal the total variance.
2. Because data are expressed in standard form in R-mode, and entities are row normalized in Q-mode, the total variance (information content) is equal to n and N, respectively. Each eigenvalue extracted accounts for a certain amount of variance; thus the percent variance explained can be calculated by dividing the eigenvalue by n (or N). By cumulating these percentages, it is

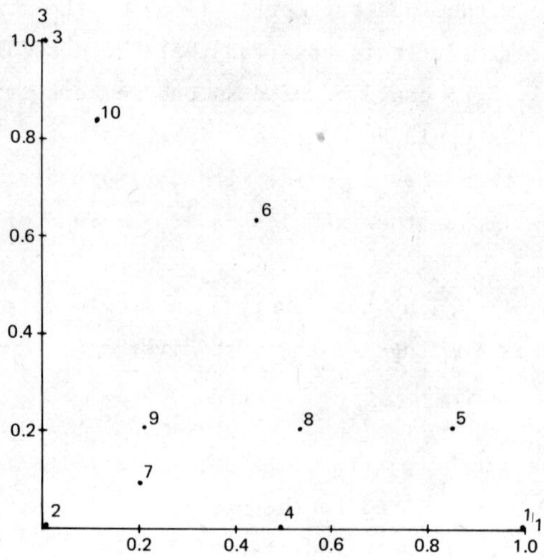

Oblique Factor Matrix

Locality	Factors 1	2	3
1	1.000	0.000	0.000
2	0.000	1.000	0.000
3	0.000	0.000	1.000
4	0.497	0.537	0.000
5	0.847	0.000	0.207
6	0.441	0.000	0.644
7	0.200	0.753	0.097
8	0.530	0.343	0.207
9	0.207	0.668	0.201
10	0.108	0.116	0.839

FIGURE 2.18

possible to arbitrarily stop extracting factors once the cumulative variance explained reaches some specified level, for example, 95%. The remaining factors needed to account for the remaining variance are assumed to represent unique factors.

3. The equation R = AA' (or S = AA') implies that the correlation matrix can be reproduced by forming the major product moment of

the factor loadings matrix. It is, of course, possible to produce an estimate of R with an n x 1 matrix A.

The difference, or residual matrix, is obtained from:

$$R_{r_1} = R - A_{n,1}A'_{1,n} \qquad (2.9.2)$$

If significant correlations remain in R_{r_1} then the second factor may be added to A and the process repeated.

$$R_{r_2} = R - A_{n,2}A'_{2,n} \qquad (2.9.3)$$

This procedure may be continued until there are no significant residual correlations. The number of columns in A is then taken as the correct number of factors.

4. The sums of squares of the factor loadings in a row of the A matrix is termed the communality and represents that proportion of the variance of a variable accounted for by the number of factors used. The "correct" number of factors can be judged by the values of communality for all the variables. If many communalities are low, say less than 0.8, more factors are probably required.
5. Because variance is extracted in descending order, factor loadings on the first few factors will be higher than those on the later ones. When all the loadings on a given factor appear to resemble nothing more than noise (error components), then this and succeeding factors may be removed from further consideration. This is often best judged on a rotated factor matrix.
6. The final criterion is entirely subjective. If factors are interpretable and "make sense" they are probably relevant. Uninterpretable factors or those whose spatial distribution forms no sensible pattern may merely represent error components.

2.10 PRACTICAL EXAMPLES

Rather than add to an already lengthy account, an annotated bibliography of a few pertinent papers is given below. Study of these papers should enable the reader to develop a better understanding of how the results of factor analysis are put to use in a variety of geologic problems.

Cameron, E. M., 1968, A geochemical profile of the Swan Hills Reef. Can. Jour. Earth Sci., v. 5, p. 287-309.

A detailed study of chemical variations in a reef complex. Factor analysis, coupled with trend surface analysis, provides a method for determining the diagenetic history of a slightly dolomitized, limestone reef.

Degens, E. T., Spencer, D. W., and Parker, R. H., 1967, Paleobiochemistry of Molluscan Shell Proteins: Comp. Biochem. Physiol., v. 20, p. 553-579.

The interrelationships between amino acids in various molluscs is studied by means of R-mode factor analysis. Environmental and genetic effects on amino acid compositions are revealed.

Harbaugh, J. W., and Demirmen, F., 1964, Application of factor analysis to petrographic variations of Americus limestone (Lower Permian), Kansas and Oklahoma: Kan. Geol. Survey Dist. Pub. 15.

A paleoecologic analysis of a thin limestone unit based on petrographic and chemical attributes. Uses both correlation coefficients and distance coefficients as similarity indices in a Q-mode analysis.

Hitchon, B., Billings, G. K., and Klovan, J. E., 1971, Geochemistry and origin of formation waters in the western Canada sedimentary basin - III. Factors controlling chemical composition: Geochim. et Cosmochim. Acta, v. 35, p. 567-598.

R- and Q-mode analyses are used to document flow paths and the chemical reactions responsible for variations in the chemistry of subsurface formation waters. Oblique rotation is used to achieve simple structure. R-mode factor scores are used as input variables to second-order factor analyses.

Imbrie, J., and van Andel, T. H., 1964, Vector analysis of heavy-mineral data: Geol. Soc. Amer. Bull., v. 75, p. 1131-1156.

The classic paper on the use of Q-mode factor analysis in the study of sediments. The Q-mode model is developed and applied to two recent sedimentary basins in a way that clearly shows the utility and power of the method.

Imbrie, J., and Kipp, N. G., 1971, A new micropaleontological method for quantitative paleoclimatology: Application to a late Pleistocene Caribbean core, in The Late Cenozoic glacial ages (Turekian, K., ed.): New Haven, Conn., Yale Univ. Press.

An involved and elegant method of analysis applied to Recent and Pleistocene foraminifera shows how the results of Q-mode factor analysis can be used in predictive, nonlinear regression models. Pleistocene oceanic temperatures and salinites can be accurately predicted on the basis of foram assemblages.

Klovan, J. E., 1966, The use of factor analysis in determining depositional environments from grain-size distributions: Jour. Sed. Petrology, v. 36, no. 1, p. 115-125.

Q-mode analysis is used to classify recent sediment samples on the basis of their grain-size distributions. The factors extracted are claimed to reflect different types of depositional energy.

Lonka, A., 1967, Trace-elements in the Finnish Precambrian phyllites as indicators of salinity at the time of sedimentation: Bull. Comm. Geol. Finlande No. 209.

Trace element variation as revealed by factor analysis leads to the surprising conclusion that depositional salinities can be determined even in highly metamorphosed shales.

Matalas, N. C., and Reiher, B. J., 1967, Some comments on the use of factor analysis: Water Resources Research, v. 3, no. 1, p. 213-223.

Some critical comments on the use and abuse of factor analysis as applied to hydrologic problems. Many mathematical and substantiative arguments are presented that must be taken into account when interpreting results. Although many of the comments are germane, others are open to serious question.

McCammon, R. B., 1966, Principal component analysis and its application in large-scale correlation studies: Jour. Geol., v. 74, no. 5, pt. 2, p. 721-733.

Explains the use of R-mode analysis as applied to crude oil variations and biostratigraphic problems. A "minimum entropy" criterion is used to achieve simple structure in rotation.

McCammon, R. B., 1969, Multivariate methods in geology, in Models of geological processes (Fenner, P., ed.): Washington, D.C., Am. Geol. Inst.

One of the best treatments of the mathematics and concepts of factor analysis and many other related topics. Algebraic and geometrical explanations are presented at an elementary level and the use of several examples makes understanding especially easy.

McElroy, M. N., and Kaesler, R. L., 1965, Application of factor analysis to the Upper Cambrian Reagan Sandstone of central and northwest Kansas: The Compass, v. 42, no. 3, p. 188-201.

An application of factor analysis to a typical stratigraphic problem. Factors are interpreted in terms of regional influences that affect thickness, grain-size, and mineralogy of a sandstone unit.

Spencer, D., 1966, Factors affecting element distributions in a Silurian graptolite band: Chem. Geol., v. 1, p. 221-249.

R-mode analysis is used to determine the underlying casual influences affecting the chemical variability of a thin shale unit. A very enlightening discussion of how the factor matrices can be interpreted is especially useful.

REFERENCES

Gower, J. C., 1967, Multivariate analysis and multidimensional geometry: The Statistician, v. 17, no. 1, p. 13-28.

Harbaugh, J. W., 1964, BALGOL programs for calculation of distance coefficients and correlation coefficients using an IBM 7090 computer: Kansas Geol. Survey Sp. Dist. Pub. 9.

Harman, H. H., 1960, Modern factor analysis: Chicago, Illinois, Univ. of Chicago Press, 471 p.

Imbrie, J. and Purdy, E. G., 1962, Classification of modern Bahamian carbonate sediments, in Classification of carbonate rocks - a symposium, Mem. 1, Amer. Assoc. Petroleum Geol., p. 253-272.

Imbrie, J., 1963, Factor and vector analysis programs for analyzing geologic data: U.S. Office of Naval Research, Tech. Rept. 6, 83 p.

Kaiser, H. F., 1958, The varimax criterion for analytic rotation in factor analysis: Psychometrika, v. 23, p. 187-200.

Klovan, J. E. and Imbrie, J., 1971, An algorithm and FORTRAN IV program for large-scale P-mode factor analysis and calculation of factor scores: Jour. Inter. Assn. Math. Geol., v. 3, p. 61-67.

McIntyre, D. B., 1969, Introduction to the study of data matrices, in Models of geological processes (Fenner, P., ed.): Washington, D.C., Amer. Geol. Inst.

Pearson, K., 1901, On lines and planes of closest fit to systems of points in space: Phil. Mag., v. 6, p. 559-572.

Spearman, C., 1904, General intelligence, objectively determined and measured: Amer. Jour. Psychol, v. 15, p. 201-293.

Thurston, L. L., 1947, Multiple factor analysis: Chicago, Illinois, Univ. of Chicago Press, 535 p.

APPENDIX 1. A PRIMER ON MATRIX ALGEBRA

1. A matrix is a rectangular chart of numbers. A matrix is symbolized by a capital letter and its size is shown by two subscripts; the first referring to the number of rows, the second the number of columns. Thus $A_{r,c}$ represents the matrix A with r rows and c columns. Any number in the matrix is termed an element. Thus, $a_{i,j}$ is the element of A in the i-th row and j-th column. Two matrices are said to be equal if all elements correspond exactly. That is, A = B if $a_{ij} = b_{ij}$ for all i and j.
2. The transpose of a matrix is another matrix in which the rows and columns are interchanged. It is symbolized by an apostrophe. Thus A' is the transpose of matrix A.
3. Special types of matrices include:
   (a) Rectangular matrix. Has more rows than columns or vice versa.
   (b) Square matrix. Has the same number of rows as columns.
   (c) Square symmetric matrix. A square matrix such that $a_{ij} = a_{ji}$ for all values of i and j. The lower left triangular part of the matrix below the diagonal is a mirror image of the upper right triangular portion.
   (d) Diagonal matrix. Only the elements in the principal diagonal are nonzero and all other elements are zero. That is $a_{ij} \neq 0$ when i = j but $a_{ij} = 0$ when $i \neq j$.
   (e) Identity matrix. A diagonal matrix whose diagonal elements all equal 1.
   (f) Column vector. A matrix with n rows but only one column.
   (g) Row vector. A matrix with n columns but only one row.
   (h) Scalar. A matrix with one row and one column.
4. Matrices can be added together (or subtracted) only if they are size compatible, that is, each matrix must have the same number of rows and columns.

   Addition or subtraction is done on an element by element basis. Thus the elements of C in C = A + B are equal to the sums of corresponding elements of A and B; $c_{ij} = a_{ij} + b_{ij}$ for all i and j.

1. Matrix Definition

$$A = \begin{bmatrix} 4 & 3 & 1 \\ 7 & 2 & 4 \\ 1 & 5 & 9 \\ 2 & 8 & 6 \end{bmatrix}$$

Size of A is 4 by 3; A_{43}

$a_{32} = 5$

2. Transpose

$$A' = \begin{bmatrix} 4 & 7 & 1 & 2 \\ 3 & 2 & 5 & 8 \\ 1 & 4 & 9 & 6 \end{bmatrix}$$

A' is the transpose of A above.

3. Types of Matrices

$$\begin{bmatrix} 5 & 1 \\ 8 & 6 \\ 0 & 2 \end{bmatrix}$$
rectangular

$$\begin{bmatrix} 5 & 3 & 8 \\ 1 & 2 & 4 \\ 7 & 0 & 2 \end{bmatrix}$$
square

$$\begin{bmatrix} 1 & 5 & 3 \\ 5 & 2 & 4 \\ 3 & 4 & 7 \end{bmatrix}$$
square symmetric

$$\begin{bmatrix} 2 & 0 & 0 \\ 0 & 5 & 0 \\ 0 & 0 & 1 \end{bmatrix}$$
diagonal

$$\begin{bmatrix} 1 & 0 & 0 & 0 \\ 0 & 1 & 0 & 0 \\ 0 & 0 & 1 & 0 \\ 0 & 0 & 0 & 1 \end{bmatrix}$$
identity

$$\begin{bmatrix} 1 \\ 3 \\ 0 \\ 4 \\ 1 \end{bmatrix}$$
column vector

$$\begin{bmatrix} 3 & 5 & 8 & 7 & 1 \end{bmatrix}$$
row vector

4. Matrix Addition

C = A + B

$$\underset{A}{\begin{bmatrix} 5 & 3 & 1 \\ 0 & 2 & 5 \\ 7 & 8 & 1 \\ 4 & 3 & 6 \end{bmatrix}} + \underset{B}{\begin{bmatrix} 2 & 6 & 0 \\ 1 & 4 & 1 \\ 2 & 0 & 3 \\ 4 & 3 & 6 \end{bmatrix}} = \underset{C}{\begin{bmatrix} 7 & 9 & 1 \\ 1 & 6 & 6 \\ 9 & 8 & 4 \\ 8 & 6 & 12 \end{bmatrix}}$$

5. Multiplication of matrices can only be performed if the number of columns of the pre-factor is equal to the number of rows of the post-factor. That is, $C = A_{n,m} \cdot B_{e,f}$ is only possible if $m = e$.

   An element of C is defined as follows:

$$c_{ij} = \sum_{k=1}^{m} a_{ik} \cdot b_{kj}$$

   where m is the number of columns of A and the number of rows of B.

   A row of C is produced by multiplying the first row of A times each column of B. This is repeated for every row of A until the C matrix is complete.

   The minor product moment is defined as $C = A'A$. C contains the sums of squares and cross products of the rows of A.

6. The trace of a square matrix is the sum of its diagonal elements.
7. The matrix analog of scalar division is accomplished by inversion. If A is a square matrix and $A \cdot B = B \cdot A = I$ then B is said to be the inverse of A. The notation A^{-1} is commonly used to denote the inverse of A.

   Finding the inverse of a matrix is a rather complicated procedure and the reader is referred to any good text on matrix algebra for details.

5. <u>Matrix Multiplication</u> $C = A \cdot B$

$$\begin{array}{cc} & B \\ & \begin{bmatrix} 2 & 1 & 0 \\ 4 & 2 & 3 \end{bmatrix} \\ \begin{bmatrix} 5 & 1 \\ 0 & 2 \\ 4 & 3 \\ 1 & 1 \end{bmatrix} & \begin{bmatrix} 14 & 7 & 3 \\ 8 & 4 & 6 \\ 20 & 10 & 9 \\ 6 & 3 & 3 \end{bmatrix} \\ A & C \end{array}$$

$$c_{11} = a_{11} \cdot b_{11} + a_{12} \cdot b_{21}$$
$$14 = 5 \cdot 2 + 1 \cdot 4$$
$$c_{12} = a_{11} + b_{12} + a_{12} \cdot b_{22}$$
$$7 = 5 \cdot 1 + 1 \cdot 2$$

In general, the following "box" notation for matrix multiplication will be found useful.

$$\begin{array}{cc} & [\text{post-factor}] \\ [\text{pre-factor}] & [\text{product}] \end{array}$$

$$\begin{array}{cc} & A \\ & \begin{bmatrix} 2 & 0 \\ 1 & 3 \\ 3 & 1 \end{bmatrix} \\ A' & \\ \begin{bmatrix} 2 & 1 & 3 \\ 0 & 3 & 1 \end{bmatrix} & \begin{bmatrix} 14 & 6 \\ 6 & 10 \end{bmatrix} \end{array} \qquad \begin{array}{cc} & \begin{bmatrix} 2 & 1 & 3 \\ 0 & 3 & 1 \end{bmatrix} \\ \begin{bmatrix} 2 & 0 \\ 1 & 3 \\ 3 & 1 \end{bmatrix} & \begin{bmatrix} 4 & 2 & 6 \\ 2 & 10 & 6 \\ 6 & 6 & 10 \end{bmatrix} \end{array}$$

minor product moment major product moment

6. Note that the trace of both products is equal to 24.

7. <u>Matrix inversion</u>. Given the square matrix A, it is necessary to find a matrix B such that AB = I or

$$\begin{array}{cc} & [B] \\ [A] & [I] \end{array}$$

Expanding the above yields

$$a_{11}b_{11} + a_{12}b_{21} + \cdots + a_{1n}b_{n1} = 1.0$$
$$a_{11}b_{12} + a_{12}b_{22} + \cdots + a_{1n}b_{n2} = 0$$
$$a_{21}b_{11} + a_{22}n_{21} + \cdots + a_{2n}b_{n1} = 0$$
$$a_{21}b_{22} + a_{22}b_{22} + \cdots + a_{2n}b_{n2} = 1.0$$

etc.

When b's are determined so as to satisfy this set of simultaneous equations then $B = A^{-1}$, the inverse of A.

8. A square matrix Q is said to be orthogonal if

$$Q'Q = D$$

where D is a diagonal matrix.

A square matrix Q is said to be orthonormal if

$$Q'Q = QQ' = I$$

9. The Rank of a Matrix. The rank of a matrix may be defined, in terms of its column or row vectors, as the number of linearly independent row, or column, vectors present in the matrix.

Another view of rank is as follows. A matrix X_{nm} can be expressed as the product of two matrices whose common order is r. If X cannot be expressed as the product of any pair of matrices with a common order less than r, then the rank of X is r. It can be appreciated from this that a very large matrix may have a low rank and thus be expressable as the product of two smaller matrices.

This is the basis of practically all multivariate methods of data analysis.

8.

$$Q = \begin{bmatrix} 2 & 3 \\ 1 & -6 \end{bmatrix} \quad \text{then } Q'Q = \begin{bmatrix} 2 & 1 \\ 3 & -6 \end{bmatrix} \begin{bmatrix} 2 & 3 \\ 1 & -6 \end{bmatrix} = \begin{bmatrix} 5 & 0 \\ 0 & 45 \end{bmatrix}$$

Q is orthogonal

$$Q = \begin{bmatrix} 0.5 & -0.866 \\ 0.866 & 0.5 \end{bmatrix} \qquad Q'Q = QQ' = I$$

Q is orthonormal

9. Rank of a Matrix

$$A = \begin{bmatrix} 6 & 3 & 12 \\ 8 & 4 & 16 \\ 12 & 6 & 24 \end{bmatrix}$$

The rank of A is 1 for there is only one linearly independent vector; all columns (or rows) are multiples of each other.

A can be reproduced by the product of an infinite number of pairs of vectors, for example:

$$\begin{bmatrix} 3 \\ 4 \\ 6 \end{bmatrix} \begin{bmatrix} 2 & 1 & 4 \end{bmatrix} = \begin{bmatrix} 6 & 3 & 12 \\ 8 & 4 & 16 \\ 12 & 6 & 24 \end{bmatrix}$$

$$A = \begin{bmatrix} 2 & 10 & 5 \\ 1 & 5 & 3 \\ 5 & 25 & 15 \end{bmatrix}$$

The rank of A is 2 for there are two linearly independent vectors; columns 1 and 2 are multiples of each other.

$$A = \begin{bmatrix} 2 & 10 & 5 \\ 1 & 5 & 3 \\ 5 & 20 & 15 \end{bmatrix}$$

The rank of A is 3.

10. Eigenvalues and Eigenvectors

Given a real square symmetric matrix A, there exist scalars λ and vectors u such that

$$Au = \lambda u$$

or

$$Au - \lambda u = 0$$

or

$$(A = \lambda I)u = 0$$

$$\begin{bmatrix} [A] - [\lambda I] \end{bmatrix} \quad \overset{[U]}{\begin{bmatrix} 0 \\ 0 \\ 0 \\ 0 \end{bmatrix}}$$

An nth-order symmetric matrix A has eigenvalues $\lambda_1, \lambda_2 \ldots \lambda_n$ possibly not all distinct and possibly some being zero. Associated with each λ is a vector (eigenvector) $u_1, u_2 \ldots u_n$, such that $u'_i u_j = 0$ for all i and j when $i \neq j$. $u'_i u_i = 1$ for $i = 1,n$. Placing the eigenvalues in a diagonal matrix Λ and the eigenvectors into Q we obtain

$$AQ = Q\Lambda$$

or

$$Q'AQ = \Lambda$$

or

$$A = Q\Lambda A'$$

which is referred to as the basic structure of the square symmetric matrix A. The rank of A is equal to the number of nonzero eigenvalues.

If there are m nonzero eigenvalues, the basic structure suggests that two small matrices contain the same information as does A, viz

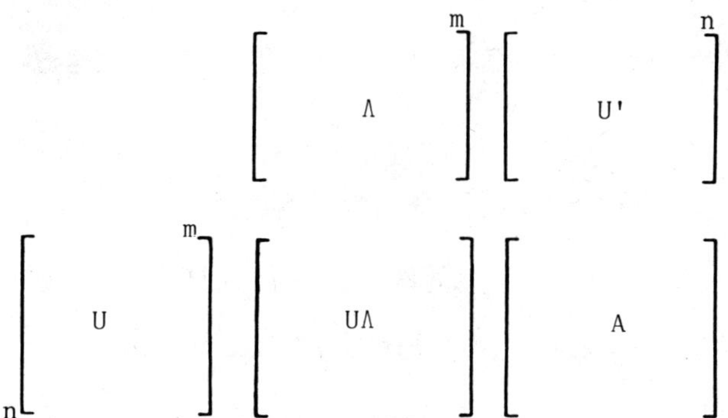
m
n
Λ
U'
m
U
UΛ
A
n

Chapter 3

Some Practical Aspects of Time Series Analysis

William T. Fox

3.1 GENERALITIES

In many geologic applications, geologic observations in a stratigraphic succession correspond to changes taking place through time. Where deposition was continuous and the rate of deposition was relatively constant, stratigraphic thickness can be considered directly proportional to time. Therefore, time-trend curves can be plotted with stratigraphic thickness corresponding to time as the independent variable and the parameter being studied as the dependent variable. The dependent variables can include such things as grain size, carbonate content, color, or fossil distribution. In modern sedimentary studies where processes are being studied through time, absolute time can be plotted as the independent variable and the dependent variables can include such things as barometric pressure, wind velocity, wave height, or current velocity.

As pointed out by Kendall (1948, pp. 363-437), a sequence of observations made in a time series are influenced by three separate components: (1) a trend or long-term component, (2) cyclical or oscillating functions about the trend, and (3) a random or irregular component. The trend is considered a broad, smooth undulating motion of the system over a relatively long period of time or through a relatively large number of sedimentation units. The cyclical or oscillating fluctuations about the trend represent a "seasonal effect" or local variations that are superimposed on the trend component. When the cyclical fluctuations and the trend have been subtracted from the data, we are left with the random fluctuations that are referred to as the random error or residuals.

Several techniques are available for separating the trend component from the oscillating fluctuations and random variations in a time series. The most straightforward method is to draw a purely interpretive curve through the clusters of high and low values on the observed data curve. This "eye balling" technique has been used effectively by Walpole and Carozzi (1961) in their study of the microfacies of the Rundel group. For illustrative purposes, a free-hand trace of the main trend is useful, but since it is highly interpretive, it would be difficult for another worker to reproduce. Also, with a free-hand method, it is difficult to separate a major trend from minor oscillations in the data.

Weiss et al. (1965) constructed smooth curves or "graphic logs" by grouping layers into units that were each 3 feet thick. The resulting "moving total" curves were used for correlating between adjacent measured sections and interpreting the depositional environment. The moving total curves gave a good picture of the gross lithologic changes but would be difficult to use for detailed stratigraphic studies. This method gave a reproducible curve for the major trend, but the small-scale oscillations or other higher-frequency components superimposed on the random fluctuations were lost.

One of the most frequently used methods for smoothing a time series is the simple moving average described by Krumbein and Pettijohn (1938, p. 198) and used by Walker and Sutton (1967, p. 1014). With this technique, the data are arranged in a stratigraphic sequence with values recorded at equal increments of time or stratigraphic thickness. Starting at the base of the section, a series of moving averages is taken on successive groups involving an odd number of data points. In practice, the successive averages are computed for a series by dropping the lowest data point and adding the next value in the sequence. As pointed out by Miller and Kahn (1962, p. 355), the moving average technique should be regarded as descriptive rather than analytical. Because the weight of each value is equally distributed within the group being averaged, the moving average technique subdues highs and low and displaces peaks and valleys in the trend. As with the previous techniques described, this method gives an approximation of the major trend, but since the highs and lows in the trend are markedly reduced, the oscillating fluctuations are exaggerated.

The three techniques that provide the most useful methods for smoothing curves include polynomial curve fitting, iterated moving averages, and Fourier analysis. Polynomial curve fitting and iterated moving averages make use of summation equations that are readily adaptable to computer programming. In Fourier analysis, a series of sine and cosine curves representing fundamental harmonics are fit to the observed data.

3.2 POLYNOMIAL CURVE FITTING

It is implicit in the concept of time trend analysis that the movement be relatively smooth over long periods of time (Kendall, 1948, p. 371). Therefore, the trend component (U_1) can be represented by a polynomial in the time element, t, as follows:

$$U_t = a_0 + a_1 t + a_2 t^2 + \cdots + a_p t^p$$

By increasing the size of p, we can obtain as close an approximation to a finite series as we desire. When the polynomial is fitted to the whole time series by the method of least squares, it gives a curvilinear regression line of U_t on the variable t. The method of fitting a polynomial to the data by least-squares analysis has worked quite successfully for trend surface analysis of facies maps. In fitting a polynomial response surface to a map, the most success has been with the linear, quadratic, and cubic surfaces. In using a polynomial fitted over the entire time series by least squares, it would be necessary to use a much higher-order polynomial to fit the trend. As pointed out by Kendall (1948, p. 371), the high-order polynomial would be somewhat artificial, and the coefficients, being based on high-order moments, would be very unstable from a sampling point of view. It would also be difficult to separate the trend component from the oscillating and random components when using a high-order polynomial.

As a possible alternative for finding a single high-order polynomial which approximates the entire time series, Kendall (1948, p. 372) suggests using a sequence of low-order polynomials representing overlapping segments of the series. In using this technique, data points must be spaced at equal intervals along the time line. The first step is to take an odd number of data points (2m + 1), with m representing the number of points on each side of the value being smoothed, and to fit a polynomial of order p, with p not greater than 2m, to them. The value of the polynomial at the middle of its range is substituted for the corresponding observed data point in plotting the smoothed curve. The polynomial fitting operation is repeated for consecutive sets of 2m + 1 terms from the beginning to the end of the time series. The degree of smoothing of the trend curve is controlled by the number of terms included in the polynomial.

In a series of 2m + 1 terms, the terms are denoted by

$$U_{-m},\ U_{-(m-1)},\ U_{-(m-2)},\ \ldots,\ U_0,\ \ldots,\ U_{m-2},\ U_{m-1},\ U_m$$

According to Kendall (1948, p. 372), the coefficients of a polynomial of the order p are obtained by the method of least squares giving an equation of the following form:

$$A_0 = C_0 + C_1U_{-m} + C_2U_{-(m-1)} + C_3U_{-(m-2)} + \cdots + C_{m+1}U_0 + \cdots + C_{2m+1}U_m \qquad (3.2.1)$$

In equation (3.2.1), the constants, C's, depend on m, the number of terms, and p, the order of the polynomial, but are independent of the U's. A_0 is equivalent to U_0 at t = 0, so this is the value for the middle of the range of the polynomial. As can be seen from equation (3.2.1), this is equivalent to a weighted average in which the weights are independent of the observed values. Therefore, to compute the trend line, the constants for equation (3.2.1) are determined for the selected values of m and p, and then the value for A_0 given in equation (3.2.1) is calculated for each consecutive set of 2m + 1 terms in the series. It should be noted that there will be a loss of m terms at the beginning and at the end of the trend curve. Tables listing the formulas for fitting a polynomial of orders 2 and 3 (quadratic and cubic) and orders 4 and 5 (quartic and quintic) for $m = 1$ to 10 are given in Kendall (1948, p. 374) and Whittaker and Robinson (1929, p. 295). The same value is obtained by fitting a polynomial of order 2 (quadratic) or order 3 (cubic) since the case p odd includes the next lowest (even) value of p. Therefore, it is not necessary to give separate values for the even (quadratic and quartic) polynomials if the odd (cubic and quintic) polynomials have been calculated (Kendall, 1948, p. 373). Four of the equations fitting a quadratic and cubic when $m = 2$, $m = 3$, $m = 4$ and $m = 5$ are given as equations (3.2.2) to (3.2.5) (Whittaker and Robinson, 1929, p. 295).

$$m = 2: \quad U'_0 = \frac{1}{35}[17U_0 + 12(U_1 + U_{-1}) - 3(U_2 + U_{-2})] \tag{3.2.2}$$

$$m = 3: \quad U'_0 = \frac{1}{21}[7U_0 + 6(U_1 + U_{-1}) + 3(U_2 + U_{-2}) - 2(U_3 + U_{-3})] \tag{3.2.3}$$

$$m = 4: \quad U'_0 = \frac{1}{231}[59U_0 + 54(U_1 + U_{-1}) + 39(U_2 + U_{-2}) + 14(U_3 + U_{-3}) - 21(U_4 + U_{-4})] \tag{3.2.4}$$

$$m = 5: \quad U'_0 = \frac{1}{429}[89U_0 + 84(U_1 + U_{-1}) + 69(U_2 + U_{-2}) + 44(U_3 + U_{-3}) + 9(U_4 + U_{-4}) - 36(U_5 + U_{-5})] \tag{3.2.5}$$

As the number of terms is increased, sizes of the constants become quite large and in moving from quadratic and cubic to quartic and quintic, the sizes of the constants also greatly increase. Because of the labor involved in their use, it is advisable to use a computer for plotting trend curves.

3.3 ITERATED MOVING AVERAGES

Several different techniques have been proposed to simplify the computations for fitting a trend line by moving averages. The iterated averages method most widely used was introduced by actuaries for "graduating" a life expectancy curve, which is similar to fitting a trend line in geology. Whittaker and Robinson (1929, p. 286) point out one of the earliest examples of iterated moving averages, which was developed by Woolhouse in 1870. Woolhouse computed each point in the trend line by passing five parabolas through the following five sets of points with three points to a set:

$$(U_{-7}, U_{-2}, U_3),\ (U_{-6}, U_{-1}, U_4),\ (U_{-5}, U_0, U_{-5}),\ (U_{-4}, U_1, U_6),$$

$$(U_{-3}, U_2, U_7)$$

To compute the graduated value, Woolhouse took the arithmetic mean of the values of the five parabolas as they passed through a line perpendicular to the time series through U_0. The values at U_0 can be determined for each of the parabolas using the Newton-Gauss formula of interpolation given in Whittaker and Robinson (1929, p. 36). The arithmetic mean of the interpolated values of U_0 can also be found by the following summation formula, which is given as equation (3.3.1):

$$U_0' = \frac{1}{125}[25U_0 + 24(U_1 + U_{-1}) + 21(U_2 + U_{-2}) + 7(U_3 + U_{-3}) + 3(U_4 + U_{-4}) - 2(U_6 + U_{-6}) - 3(U_7 + U_{-7})] \qquad (3.3.1)$$

The Woolhouse 15-term formula using seven terms on each side of the central value has about the same degree of smoothing as the nine-term formula given as equation (3.2.3). The Woolhouse formula using iterated moving averages gives a smoother trend curve than the fitting of a quadratic or cubic polynomial to the data.

Another type of iterated moving average formula using three successive averages covering 15 points was developed by Spencer and is given by Kendall (1948, p. 376). The first moving average used for five terms has a constants -3, 3, 4, 3, -3. The values resulting from this moving average are averaged first in sets of five points each, then these values are averaged twice in sets of four. The form used for such an iteration is given as

$$U_0' = \frac{1}{320}[4]^2[5]\ [-3, 3, 4, 3, -3] \qquad (3.3.2)$$

When the separate iterations are combined into a complete summation formula, the weights are those given in equation (3.3.3), which is Spencer's 15-point formula.

$$U_0' = \frac{1}{320}[74U_0 + 67(U_1 + U_{-1}) + 46(U_2 + U_{-2}) + 21(U_3 + U_{-3}) + 3(U_4 + U_{-4}) - 5(U_5 + U_{-5}) - 6(U_6 + U_{-6}) - 3(U_7 + U_{-7})] \quad (3.3.3)$$

Expanding the same technique that was used for the Spencer 15-point formula [equation (3.3.3)], Spencer also developed a 21-point equation explained by Whittaker and Robinson (1929, p. 290). In using this formula, the seven-term series (-1,0,1,2,1,0,-1) is averaged and the values are averaged first in sets of seven, then twice in sets of five, as is shown in

$$U_0' = \frac{1}{350}[5]^2[7]\ [-1, 0, 1, 2, 1, 0, -1] \quad (3.3.4)$$

Spencer's 21-term formula can be expanded into the following summation formula;

$$U_0' = \frac{1}{350}[60\ U_0 + 57(U_1 + U_{-1}) + 47(U_2 + U_{-2}) + 33(U_3 + U_{-3}) + 18(U_4 + U_{-4}) + 6(U_5 + U_{-5}) - 2(U_6 + U_{-6}) - 5(U_7 + U_{-7}) - 5(U_8 + U_{-8}) - 3(U_9 + U_{-9}) - (U_{10} + U_{-10})] \quad (3.3.5)$$

When the computations must be done by hand or with a desk calculator, it is useful to set up a table to carry out the successive averaging. Vistelius (1961) used such a table to compute the trend terms for Spencer's 21-term formula. By using cardboard cutouts exposing only portions of the table, it is possible to compute about 600 graduated values per day. In using the table with Spencer's 21-term equation, the first step is to form the computation $(1/2)(-U_3 + U_1 + 2U_0 + U_{-1} - U_{-3})$ for the entire series. The values derived from the first computation are then summed by sevens and divided by seven, then summed twice by fives, dividing by five each time. The actual order in which the iterations are carried out is immaterial, but with a long series it is advisable to do the more complicated operations while the numbers are still small.

The "goodness of fit" of a smoothed curve to the original curve may be expressed as the percentage reduction in the total sum of squares, which is given by the expression:

$$100 \frac{\Sigma x^2_{trend} - \frac{(\Sigma x_{trend})^2}{n}}{\Sigma x^2_{obs} - \frac{(\Sigma x_{obs})^2}{n}}$$

where

x_{trend} = values on trend surface at location of data points
x_{obs} = observed data values
n = number of data values

Obviously, a perfect fit of the curve to the data points would give 100 percent and any less perfect fit would yield a correspondingly smaller percentage of total sum of squares.

Nine smoothing equations are available with the (Fox, 1964) program for computing and plotting trend curves with varying degrees of smoothing. Formulas derived by Sheppard (Whittaker and Robinson, 1929, p. 279) for fitting a quadratic and cubic polynomial to m points, with m varying from 2 to 10, are used in the program. Equations (3.2.2) to (3.2.5) in this chapter are the first four equations used with the program. By increasing the number of terms in the smoothing equation, the fluctuations in the data are subdued and the underlying trends of sedimentation are accentuated. As a characteristic of the type of smoothing equation, there are still minor fluctuations in the data, even when the 21-term equation is used for smoothing fossil data (Figure 3.1). When using an iterated moving average as done by Woolhouse's 15-term equation and Spencer's 15- and 21-term equations, the minor fluctuations are completely removed, leaving only the smooth trend curve. Since the program was published (Fox, 1964), it has been modified (Fox, 1968) with the addition of the Woolhouse 15-term equation of equation (3.2.5) and the Spencer 21-term equation of equation (3.3.5). In degree of smoothing, the Woolhouse equation is equivalent to the Sheppard nine-term equation (m = 4), and the Spencer 21-term

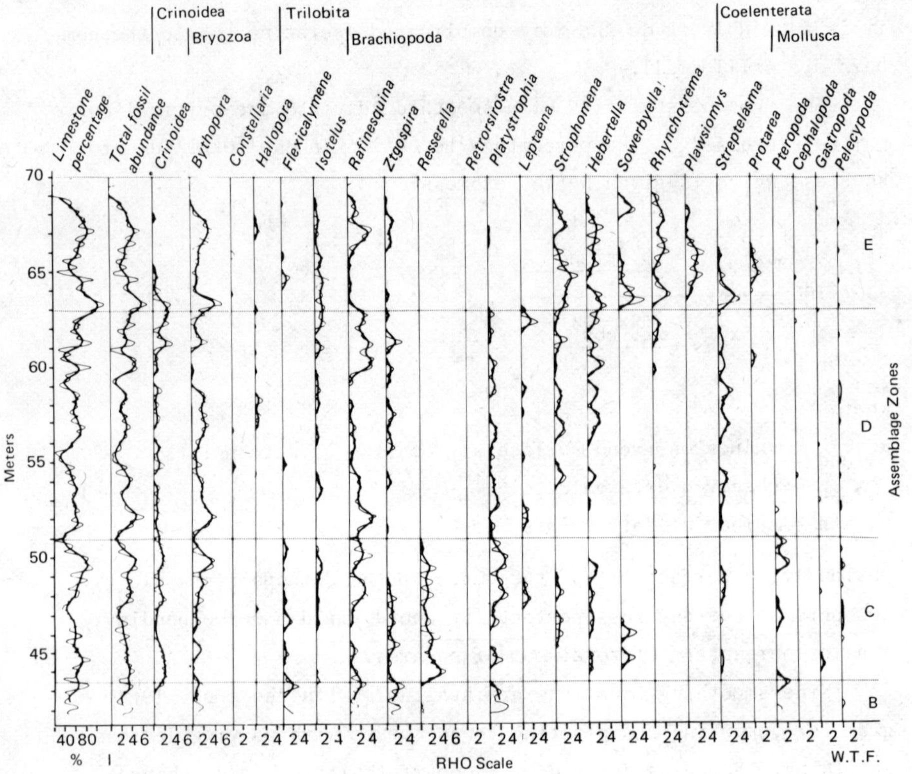

FIGURE 3.1

equation is equivalent to the Sheppard 11-term equation (m = 5). Because of the more uniform smoothing, the iterated moving average curves are preferred over the polynomial fitting curves. The only apparent disadvantage to using the Spencer and Woolhouse formulas is the loss of points at the beginning and end of the curve. Since the series being smoothed is quite long relative to the loss at each end, the overall effect is not too bad.

3.4 FOURIER ANALYSIS

Geologic processes that are cyclic in nature can be best described using Fourier analysis. In Fourier analysis, a complicated curve can be broken down into an aggregate of simple wave forms described by a series of sine and cosine curves. The observed data can be expressed as a series of fundamental harmonics that are theoretically independent. Each harmonic has a wavelength that is a discrete fraction of the total observation period. For each harmonic, the wavelength is defined as the distance from crest to crest and the amplitude as one-half the height from trough to crest. In Fourier analysis of geologic processes, wavelength can be expressed in time or stratigraphic thickness and the amplitude in the observed units for each parameter.

The complicated form of the observed data can be represented by an aggregate of simple wave forms that are expressed by the amplitude of the cosine and sine terms, a_n and b_n, respectively. Although the function of the form $Z = f(x)$ is not known, data points (x_i, z_i), are available at equal intervals. Thus, the coefficients a_n and b_n may be determined by numerical integration methods employing equations (3.4.1) and (3.4.2) and used in equation (3.4.3) to approximate the observed curve according to methods described by Harbaugh and Merriam (1968) and Fox and Davis (1971).

$$a_n = \frac{2}{K}\left[\frac{Z_0 + Z_K}{2} + \sum_{i=1}^{K-1} z_i \cos \frac{n\pi x_i}{L}\right] \quad n = 0,1,2,\ldots,K/2 \tag{3.4.1}$$

$$b_n = \frac{2}{K}\sum_{i=1}^{K-1} Z_i \sin \frac{n\pi x_i}{L} \quad n = 1,2,\ldots,K/2 \tag{3.4.2}$$

$$Z_i = F(x_i) = \frac{a_0}{2} + \sum_{n=1}^{N} a_n \cos \frac{n\pi x_i}{L} + b_n \sin \frac{n\pi x_i}{L} \tag{3.4.3}$$

where

z_i = observed value at i-th sampling point,

$F(X_i)$ = value of approximating function at i-th sampling point

a_0 = coefficient of zeroth degree cosine term, which is equal to the mean

n = degree of term

a_n = coefficient of cosine terms, $n = 1, 2, \ldots, \infty$

b_n = coefficient of sine terms, $n = 1, 2, \ldots, \infty$

π = 3.1416

x_i = sampling point, time in this case

$i = 0, 1, 2, \ldots, K$

K = maximum number of sampling points (an even number)

L = half of fundamental sampling length $K\ \Delta x/2$

N = maximum degree of series, $N = K/2$.

In analyzing geologic processes, it is convenient to plot each harmonic as a single sine curve with a given phase and amplitude. In Fourier analysis, each harmonic is expressed by a pair of sine and cosine curves with the same period. When the sine and cosine curves are added algebraically, a new sine curve results with a phase shift and a new amplitude. The phase shift can be determined by using an arc-tangent subroutine (Louden, 1967). The phase for each harmonic, p_n, can be computed according to

$$p_n = \text{arc tan}\,\frac{a_n}{b_n} \tag{3.4.4}$$

The phase that is expressed in degrees is used to determine the starting point for a sine curve for each individual harmonic. Since the period for each harmonic is expressed in hours, it is also possible to convert the phase into hours. In this way, it is possible to compare coastal parameters such as wave height or wave period by comparing the phase shifts of different harmonics. The amplitude, σ_n, for each harmonic can be determined directly from the power spectrum (Preston and Henderson, 1964). The discrete power spectrum, σ_n^2, is defined as the sum of the square Fourier coefficients according to

$$\sigma_n^2 = a_n^2 + b_n^2 \qquad n = 0, 1, 2, \ldots, K/2 \tag{3.4.5}$$

The term "power spectrum" arose because of its relation to the power dissipation in an alternating current circuit (Harbaugh and Merriam, 1968). The amplitude for each harmonic is derived from the square root of the power spectrum according to equation (3.4.6)

$$\sigma_n = \sqrt{a_n^2 + b_n^2} \qquad n = 0, 1, 2, \ldots, K/2 \tag{3.4.6}$$

where the phase, p_n, is expressed in radians and the amplitude, σ_n, is derived from the power spectrum. The height of the curve, z_i, can be computed at each sampling point using

$$z_i = \frac{a_0}{2} + \sum_{n=1}^{N} \sigma_n \sin \frac{n\pi(p_n + x_i)}{L} \qquad (3.4.7)$$

The amplitudes of the Fourier coefficients are especially meaningful because they are in the same units as the original data. Thus, if wave height is measured in feet or longshore current velocity in feet per second, the Fourier coefficients will be expressed in the same respective units.

Since each Fourier component is a discrete harmonic of the curve for the observed data, the Fourier components are theoretically independent. The number of Fourier components available is equal to one-half the total number of data points or observations. Therefore, with 360 observations taken at 2-hour intervals over 30 days, it is possible to obtain 180 Fourier harmonics with periods ranging from 4 to 720 hours. With least squares techniques, the total variance accounted for by each harmonic can be calculated. Theoretically, a curve consisting of the full 180 Fourier harmonics should account for 100 percent of the total sum of squares. In practice, a small number of basic harmonics usually accounts for a very large percentage of the total sum of squares. Where a particular harmonic or set of harmonics are related to a naturally occurring cycle a large percentage of the sum of squares can be accounted for by a small number of harmonics.

3.5 AN APPLICATION

Barometric pressure, which is a major indication of the weather patterns passing through an area, has been selected to demonstrate how Fourier analysis is used. The observed curve for barometric pressure from 8:00 a.m., June 29 through 6:00 a.m., July 29, 1970, is plotted across the top of Figure 3.2. From the observed data, it can be seen that four major low-pressure systems passed through the area on July 4, 9, 15, and 19, 1970. Minor low-pressure systems, which cause small fluctuations in the observed curve, can be seen on July 1, 17, and 24. The major low-pressure systems are accompanied by high wind and waves that caused beach erosion or deposition, and changes in the configuration of the nearshore bars.

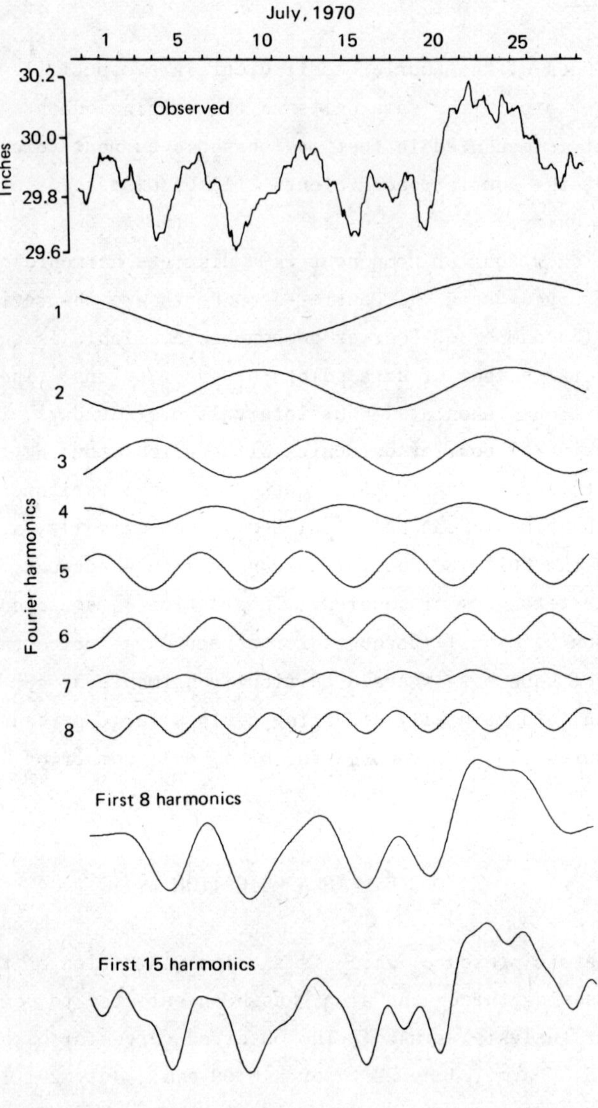

FIGURE 3.2

The period, cosine, sine, phase, amplitude, and sum of squares for the first 15 Fourier harmonics for barometric pressure are given in Table 3.1.

TABLE 3.1

Harmonic, n	Period	Cosine, a_n	Sine, b_n	Phase, p_n	Amplitude, σ_n	Sum Sq., %
1	720.0	0.059	-0.096	296.8	0.112	33.6
2	360.0	-0.024	-0.069	199.2	0.073	15.2
3	240.0	-0.053	0.046	207.2	0.070	15.1
4	180.0	0.031	0.005	40.4	0.031	2.1
5	144.0	0.045	0.049	17.0	0.067	11.1
6	120.0	-0.051	-0.033	88.7	0.051	8.8
7	102.6	-0.012	0.016	92.0	0.020	1.3
8	90.0	0.028	-0.029	34.1	0.041	4.1
9	80.0	0.012	-0.008	27.3	0.014	0.2
10	72.0	-0.009	-0.008	45.4	0.012	0.7
11	65.3	0.005	0.007	6.4	0.009	0.1
12	60.0	0.007	-0.009	23.9	0.011	0.2
13	55.2	-0.032	-0.009	38.2	0.024	2.2
14	51.3	0.020	-0.004	14.5	0.020	0.8
15	48.0	0.007	-0.019	21.2	0.020	0.9

The first harmonic has a period of 30 days or 720 hours and an amplitude of 0.112 inches of mercury. The phase for the first harmonic is 148.4 degrees or 296.8 hours. The first harmonic, which accounts for approximately 33.6 percent of the total sum of squares, is plotted as the second curve in Figure 3.2. The curve for the first harmonic has a low on July 9 and a high on July 24. This is strongly influenced by the low-pressure systems early in the month and the high-pressure system that passed through the area late in the month. The second harmonic plotted as the third curve in Figure 3.2 has a period of 15 days or 360 hours, an amplitude of 0.73 inches, and a phase of 199.2 degrees and hours. This harmonic has highs on July 10 and 25 and lows on July 2 and 17. Although the curve for the second harmonic does not appear to agree with the observed data, it accounts for 15.2 percent of the total sum of squares. The curves for the third through eighth harmonics are also plotted in Figure 3.2 along with the cumulative curve for the first eight harmonics. The first eight harmonics for barometric pressure account for 90.6 percent of the total sum of squares. As with any periodic data having a wave form, the harmonics interfere with each other resulting in the reinforcement or cancellation at certain parts of the curve. In the cumulative curve for the first eight harmonics the major high- and low-pressure systems in the observed data can be easily recognized.

In order to get a closer approximation of the mathematical function representing barometric pressure, the first 15 Fourier harmonics were computed from the observed data. The cumulative curve for the first 15 harmonics, which accounts for 95.4 percent of the total sum of squares, includes the minor lows on July 1, 17, and 24. The 15th harmonic has a period of 48 hours or two days, therefore the bottom curve in Figure 3.2 accounts for all the variation in the data which has a period of one day or longer. The residual obtained by subtracting the 15-term curve from the observed data still accounts for 4.6 percent of the total sum of squares. This can be accounted for by the diurnal variation in barometric pressure due to heating during the day and cooling off at night. The normal diurnal fluctuation of barometric pressure has an amplitude of about 0.03 inches of mercury. By using Fourier analysis, therefore, it is possible to eliminate the diurnal variation from the barometric pressure curve. It is also possible to compare barometric pressure with other environmental parameters by comparing the phase and amplitude for each of the Fourier components. By keeping the number of harmonics constant, it is possible to visually compare the computed curves. The period, phase, and amplitude for each of the environmental parameters are given in Fox and Davis (1971).

In the northern hemisphere, winds circulate in a counterclockwise direction around a low-pressure system. During the summer months, the low-pressure systems generally pass to the north of the study area located on the eastern shore of Lake Michigan. Therefore, as the low-pressure system approaches the area, winds blow out of the southwest and generate waves from that direction. As the front passes over, the wind builds up in intensity and shifts over to the northwest. Since the winds following the passage of the front are generally stronger, the waves from the northwest are higher and have a longer period. During the high wave conditions following the passage of the front, the waves run up on the beach and water percolates into the groundwater system.

The 15-term Fourier curves for wind velocity, wave period, breaker height, and groundwater table level are plotted in Figure 3.3. The period, phase, and amplitude for each of the harmonics are given in Fox and Davis (1971). Wind, which is the driving force, controls the wave period and breaker height, which in turn influence the level of the groundwater table. Therefore, a phase lag would be expected in the Fourier curves with wind velocity reaching a peak first, followed by wave period and breaker height, with the groundwater table

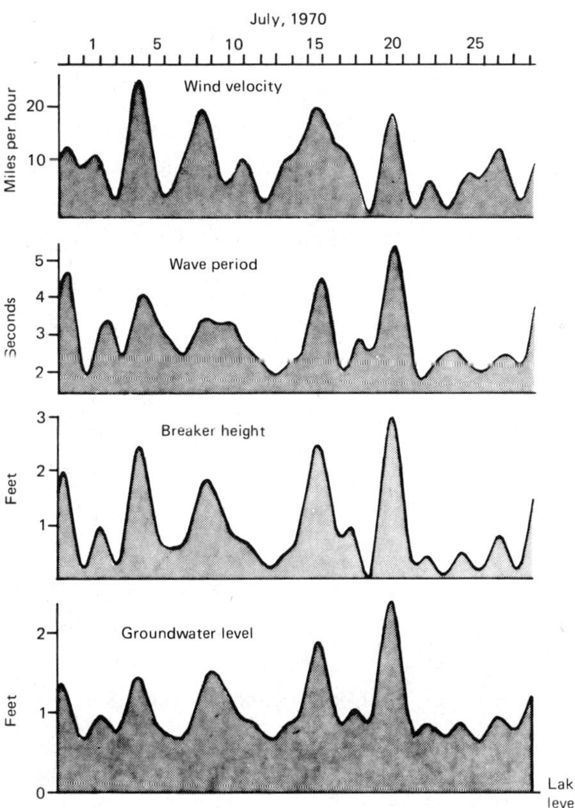

FIGURE 3.3

responding a few hours later. The 15-term Fourier curve for wind velocity shows peaks that correspond to the low points in the barometric pressure curve in Figure 3.2. The maximum wind velocity was recorded at 8:00 a.m. on July 4, 1970. The curves for wave period and breaker height in Figure 3.3 correspond quite closely to the wind velocity curve. The curve for wave period has its peaks a few hours after breaker height and drops off more slowly than breaker height. As waves change from storm waves to swells, the breaker height decreases and the wave period increases.

There is a surprisingly close correspondence between the curves for breaker height and groundwater level in Figure 3.3. Groundwater level was measured in three tubes located approximately 10, 21, and 32 feet from the plunge zone. Since the plunge zone moves with time, average distances are given for the groundwater tubes. For each of the groundwater tubes, the sixth Fourier harmonic with a period of 120 hours, or 5 days, accounts for the greatest percentage of the total sum of squares. For the first groundwater tube, the sixth harmonic has a phase of 9.8 hours. For the second tube, the same harmonic has a phase of 12.3 hours and for the third tube, it has a phase of 16.8 hours. This yields a phase difference of approximately 7 hours between the first and the third tubes. Since the tubes are 22 feet apart, this indicates that the groundwater that was fed into the foreshore by the run-up of the beaches percolates through the beach at a rate of approximately 3 feet per hour. Therefore, the curve for the second groundwater tube, which is given in Figure 3.3, has a phase lag of approximately 7 hours behind the breaker height curve.

The reversal of wind direction with the passage of low pressure systems plays an important role in controlling coastal processes. Plots of the alongshore component of the wind, longshore current velocity, and breaker angle are given in Figure 3.4. As a low-pressure system approaches the study area, wind and waves are generated from the southwest. The waves that move onto the shore from the southwest generate longshore currents that move to the north. As the low-pressure system passes over the area, wind direction shifts over to the northwest, generating waves out of that direction. With the shift in wind direction, breaker angle and longshore current are reversed

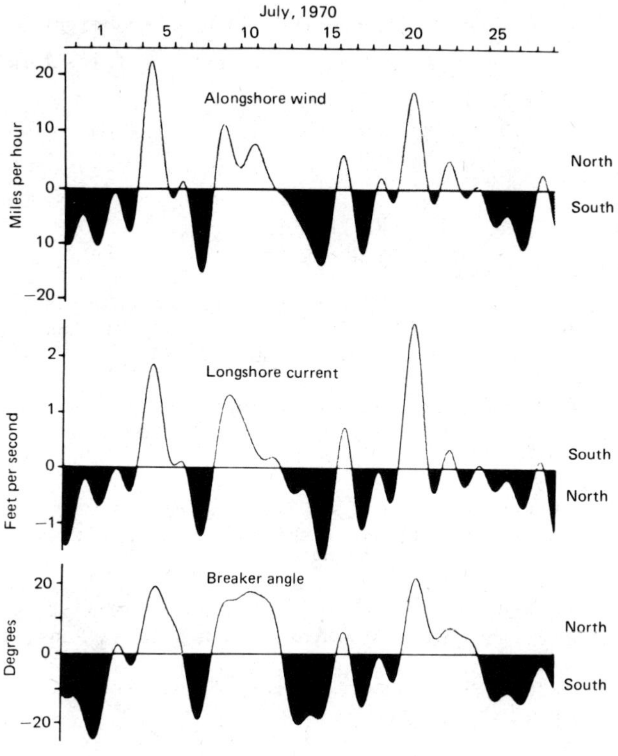

FIGURE 3.4

with the current flowing to the south. Wind and waves approaching from the northwest, accompanied by a southward flowing longshore current, are recorded as positive, while wind and waves from the southwest with a northward flowing longshore current are considered negative. As a low-pressure system approaches, a gradient wind is generated around the low-pressure system which spirals counterclockwise inward toward the center. Since the fronts pass to the north of the study area during the summer, the counterclockwise winds are blowing out of the southwest as the front moves into the sea. As the front approaches, the winds increase in velocity, building up the heights of the breakers and increasing the velocity of the longshore currents. After the front passes over, the winds shift over to the northwest followed by a corresponding shift in breaker angle and longshore current direction. The storm cycle pattern, with a low in barometric pressure accompanied by a peak in wind velocity and breaker height and a reversal in longshore current direction, is repeated several times in Figures 3.2, 3.3, and 3.4.

REFERENCES

Fox, W. T., 1964, FORTRAN and FAP program for calculating and plotting time-trend curves using an IBM 7090 or 7094/1401 computer system: Kansas Geol. Survey Spec. Dist. Pub. 12, 24.

Fox, W. T., 1968, Quantitative paleoecologic analysis of fossil communities in the Richmond Group: Jour. Geology, v. 76, pp. 613-640.

Fox, W. T. and Davis, R. A., Jr., 1971, Fourier analysis of weather and wave data from Holland, Michigan, July, 1970: O.N.R. Tech. Report No. 3, Contract 388-092, 79 p.

Harbaugh, J. W., and Merriam, D. F., 1968, Computer applications in stratigraphic analysis: New York, John Wiley & Sons, 282 p.

Kendall, M. G., 1948, The advanced theory of statistics: London, C. Griffin & Co., 503 p.

Krumbein, W. C., and Pettijohn, F. J., 1938, Manual of sedimentary petrography: New York, Appleton-Century Co., 549 p.

Louden, R. K., 1967, Programming the IBM 1130 and 1800: Englewood Cliffs, N.J., Prentice-Hall, Inc., 433 p.

Miller, R. L., and Kahn, J. S., 1962, Statistical analysis in the geological sciences: New York, John Wiley & Sons, 483 p.

Preston, F. W., and Henderson, J. H., 1964, Fourier series characterization of cyclic sediments for stratigraphic correlation, in symposium on cyclic sedimentation (Merriam, D. F., ed.): Kansas Geol. Survey Bull. 169, v. 2, pp. 415-425.

Vistelius, A. B., 1961, Sedimentation time-trend functions and their application for correlation of sedimentary deposits: Jour. Geology, v. 69, pp. 703-728.

Walker, R. G., and Sutton, R. G., 1967, Quantitative analysis of turbidites in the Upper Devonian Sonyea Group, New York: Jour. Sed. Petrology, v. 37, pp. 1012-1022.

Walpole, R. L., and Carozzi, A. V., 1961, Microfacies study of the Rundle Group (Mississippian) of Front Ranges, Central Alberta, Canada: Am. Assoc. Petroleum Geologists Bull., v. 45, pp. 1810-1846.

Weiss, M. P., Edwards, W. R., Norman, C. E., and Sharp, E. R., 1965, The American Upper Ordovician standard. VII. Stratigraphy and petrology of the Cynthiana and Eden Formations of the Ohio Valley: Geol. Soc. Amer. Spec. Paper 81, 76 p.

Whittaker, E. T., and Robinson, G., 1929, The calculus of observations in A treatise of numerical mathematics (2d ed.): London, Blackie & Sons, 395 p.

Chapter 4

Markov Models in the Earth Sciences

W. C. Krumbein

4.1 FUNDAMENTALS

The term "random process" has an unfortunate connotation for many earth scientists. It seems to imply a haphazard, unorganized, sporadic, and unpredictable process that violates the basic principles of science. These principles rest heavily on the fact that science seeks for systematic, patterned responses from recognizable causes and that unpredictable or chance events have no place in scientific analysis. Hence, by extension, models that postulate any kind of random occurrences are naturally held suspect.

Much of this misunderstanding arises from lack of recognition that a random variable has as valid a basis in scientific investigation as the conventional nonstochastic variable (systematic variable) that forms the basis of classical mathematical physics. A random variable is a mathematical entity that arises from probabilistic mechanisms, just as systematic variables are associated with deterministic mechanisms. The outcome of a deterministic experiment is exactly predictable from knowledge of the relations between dependent and independent variables, whereas the outcome of a probabilistic experiment depends on the liklihood of a given event occurring in some underlying set of probabilities. This set of probabilities constitutes the sample space of the probabilistic mechanism; if this is known, then the group behavior of the variables is completely predictable in rigorous mathematical terms. Moreover, the probability of a particular event occurring can also be exactly stated.

4.2 A SPECTRUM OF MODELS

In virtually all fields of science the range of process mechanisms extends from fully path-dependent deterministic models (in which past events completely control future events) to independent-events models, in which the past has no influence whatever on future events. A simple example of a deterministic model is the negative exponential process in time,

$$f(t) = Y = Y_0 e^{-at} \tag{4.2.1}$$

where the dependent variable Y is completely controlled by the constant Y_0, the fixed exponent a, and the independent variable t. If Y_0 and a are known either from theory or experiment, the value of Y associated with any time t can be exactly predicted. Although equation (4.2.1) is a continuous function, it can be discretized by considering successive values of f(t) at some small increment Δt. In this way the "state of the system" can be thought of in terms of discrete points in time t_{n-2}, t_{n-1}, t_n, t_{n+1}, and so on. For some phenomena the distance, X, can be substituted for time.

At the other end of the spectrum is the independent-events model. In discrete form this is expressed as

$$P\{t_{n+1} = j \mid t_n = i\} = p_j \tag{4.2.2}$$

which says that the probability of the system being in state j at time t_{n+1}, given that the state at t_n is i, is simply the probability of state j occurring at time t_{n+1}, wholly independently of the previous state of the system.

Somewhere between these extremes lie processes in which partial dependencies are present, in the sense that the state of the system at t_{n+1} does depend on the state at t_n, but is not influenced by earlier states as at t_{n-1}. This particular case gives rise to the simplest kind of Markov chain, a discrete-time, discrete-state, one-step memory process expressed as

$$P\{t_{n+1} = j \mid t_n = i\} = p_{ij} \tag{4.2.3}$$

in which p_{ij} is the <u>conditional probability</u> of the system being in state j at t_{n+1}, given that it was in state i at t_n. Here p_{ij} is the <u>transitional probability</u>: the probability that the system changes from state i to state j in the discrete time step from t_n to t_{n+1}.

4.3 THE MARKOV CHAIN

In its simplest classical form the first-order, discrete-state, discrete-time Markov chain can be visualized as representing a system with a finite number of discrete states, A, B, C, ..., behaving in such a way that a transition occurs from state i to j (where j may be either the same state or a different state) at each tick of a conceptual "Markovian Clock." The model is expressed as a transition probability matrix with rows represented by i and columns by j. A three-state system can be shown as:

$$\begin{array}{ccc} & & \underline{\text{To State } j} \\ & & \begin{array}{ccc} A & B & C \end{array} \\ \underline{\text{From State } i} & \begin{array}{c} A \\ B \\ C \end{array} & \begin{bmatrix} p_{AA} & p_{AB} & p_{AC} \\ p_{BA} & p_{BB} & p_{BC} \\ p_{CA} & p_{CB} & p_{CC} \end{bmatrix} \end{array}$$

Here, p_{AA}, p_{BB}, and p_{CC}, commonly designated as p_{ii}, represent transitions from a given state to itself, whereas the offdiagonal entries, designated as p_{ij} where $j \neq i$, represent transitions to other states. The notation in the matrix is such that, for state A, p_{AA} is the probability that the system will remain in the same state, p_{AB} is the probability that the system will move to state B at a given clock tick, and p_{AC} is the probability that it will move to state C. These three probabilities sum to 1.0. Note that when transitions occur from a given state to itself, no change in the system is apparent to an onlooker until the system, on some given clock tick, does change to a different state. The length of time that the system remains in a given state after having entered it at a particular click, is called the (discrete) waiting time for state i. Literally it refers to the number of clock ticks that the system "waits" in state i before leaving i for another state j≠i.

In structuring data for this simplest Markov chain, observations of state are made at each tick of the clock. This of course is an imaginary clock, and in practice one selects a fixed time interval based on theory, observation of what is going on, or even simple geologic intuition. In stratigraphic applications observations of state are made at fixed vertical intervals along a stratigraphic section, say at every foot. Thus, if state A represents sandstone, state B shale, and state C limestone, a sequence of such observations upward through the section might be AAABBCBBBABBCCCA..., which says that the system starts in state A and remains there for two more clock ticks, after which it changes to state B and remains there for a second tick, changes to state C for one click, then returns to state B, and so on. If observations are made at 1-foot intervals, the section has 3 feet of sandstone at the base, followed by 2 feet of shale, 1 foot of limestone, and so on.

In this particular procedure, equal increments of distance are used instead of equal increments of time. This is a matter of convenience, and although several severe geologic implications may be involved, the matrix is still expressed as transition probabilities, but now the waiting time is a discrete "thickness time." Krumbein and Dacey (1969) refer to this kind of structuring as a Markov chain with transition matrix P.

The actual procedure for assembling the transition matrix is given in Krumbein (1967, p. 3). Here, we follow through with the simplest case.

Transition probability matrices can be generated from any succession of events, but this of itself gives no indication whether the process is Markovian. Statistical tests are available for making the decision; the most widely used is that of Anderson and Goodman (1957), which tests the hypothesis of an independent-events model against the alternative that a first-order Markovian property is present.

Even if the hypothesis of an independent-events model is rejected, and that of a first-order chain accepted at least by implication, there is a second requirement that must be fulfilled. This is that the distribution of waiting times for each state must be geometrically distributed. This follows from the fact that the output from any simulation or Monte Carlo studies with Markov chains having matrix $P = [p_{ij}]$ is distributed in this manner with parameter $(1 - p_{ii})$. This requirement is so important that it deserves detailed examination.

4.4 GEOMETRIC DISTRIBUTION

This distribution is conveniently examined in terms of independent-events models of the kind shown in equation (4.2.2). Consider a single six-sided die, in which the sample space has six elements, representing the six faces, each with its pattern of dots. Each face has probability 1/6 of occurring faceup on any toss, and probability 5/6 that some other face will appear on top. If we consider the die in these terms, we have a simple system of two states. If a given face, say 4, comes up on a given toss, we can translate this into a "waiting time" by asking how long the system will remain in this state, i.e., what is the liklihood that a 4 will occur once, twice, or more times in successive throws before a non-4 shows up?

We can set up an equation for this as follows: Let $P\{R = k\}$ be the probability that the number of times, R, that a 4 comes face up is exactly k, where $k = 1, 2, 3, \ldots$. The probability that a 4-face appears initially is $p = 1/6$, and the probability that it will not appear on the next trial is $1 - p = 5/6$. Thus, once a 4-face occurs, the probability that it will occur exactly k times means that it must be repeated exactly $k - 1$ times, so that on the k-th trial some face other than a 4 will occur. This leads to the <u>geometric density with parameter (1 - p)</u>:

$$P\{R = k\} = (1 - p)\ (p)^{k-1} = (5/6)\ (1/6)^{k-1} \tag{4.4.1}$$

In this expression a "success" occurs when on the k-th trial a non-4 appears face up.

The successive probabilities are easily calculated. By setting $k = 1$, we obtain $P\{R = 1\} = (5/6)\ (1/6)^0 = 5/6 = 0.8333$. For $k = 2$

this becomes $(5/6)\ (1/6)^1 = 0.1389$; for $k = 3$ we have $(5/6)\ (1/6)^2 = 0.0232$; and the probability that k exceeds 3 is only 0.0046.

The geometric distribution applies to all independent-events models of the kind expressed by equation (4.2.2). The reason that it also applies to each state of a Markov chain with transition matrix $P = [p_{ij}]$ is that each row of the transition matrix is in fact a two-state independent-events model (a Bernoulli model) in that when the system is in state i, the probability of remaining in that state on the next tick of the clock is p_{ii} and the probability that it will change to another state $j \neq i$ is $(1 - p_{ii})$. The one-step memory of the Markov chain is thus related to the outcome of a random draw that determines whether the next drawing is to be made from the same row (state) or from some other row as specified by the offdiagonal p_{ij}'s for $j \neq i$.

The full development of the geometric distribution as it applies to the simplest Markov chain is given in Krumbein and Dacey (1969, p. 83), which includes an example with a histogram. It is interesting to note, incidentally, that $(1 - p_{ii})$ is estimated by the reciprocal of the arithmetic mean of the geometric distribution. In terms of equation (4.4.1), p_{ii} is the probability of remaining in state i for any individual drawing, and $(1 - p_{ii})$ is the probability of leaving state i on any given tick of the clock.

4.5 PROBABILITY TREES

The transition matrix P can be used to show the probabilities associated with each state of the system through a succession of ticks on the Markovian clock. Such a tree is illustrated in Krumbein (1967, p. 27) and in Harbaugh and Bonham-Carter (1970, p. 115-117). The diagrams are very instructive, and with patience one can extend the tree until succeeding sets of branches all have fixed probabilities. At this point the system has reached equilibrium, and the fixed probabilities associated with each state in the system express the overall average relative proportion of that component in the system under study.

When stratigraphic data are structured in the equal-interval form, with vertical spacing of h feet, the fixed probability vector x 100 gives the overall percentage of the total thickness of each lithology in the section. Mathematically the fixed vector is obtained by raising the matrix P to successively higher powers, and noting when all the rows of the transition matrix achieve the same values.

The fixed probability vector is actually an independent-events model, and simulations arising from it have geometric distributions as in equations (4.4.1), with parameters $(1 - p_i)$, where p_i is now the fixed probability of the i-th component.

This simplest Markov model serves to bring out some essential points in applications of Markov processes in geology. Recall that it is a discrete-state, discrete-time, one-step memory model. Variants of this model may have two or more steps in their memories (Pattison, 1965; Schwarzacher, 1967, James and Krumbein, 1969, p. 552). Moreover, the chains may be converted to continuous-time models (Krumbein, 1968a, b), and (though this becomes mathematically complex), the states may be made continuous instead of discrete.

A point not previously emphasized is that the probabilities in a Markov matrix must be stationary in the sense that the transition probabilities remain the same through the entire system being studied. Harbaugh and Bonham-Carter (1970, p. 122) develop this topic in greater detail.

A very large variety of experiments may be conducted in the framework of this simplest Markov model. Time or distance may in general be interchanged, and neither of these need to be observed at fixed intervals.

A question that emerges when Markov models are examined in detail concerns proper procedures when the input data are not geometrically distributed. This is particularly appropriate in stratigraphic analysis, where much observational data suggest that rock thicknesses are distributed lognormally rather than geometrically. We examine this situation next.

4.6 EMBEDDED MARKOV CHAINS

When a set of real-world data has the Markov property but does not display a geometric distribution for each state, an embedded Markov chain may be more appropriate for analysis. This version of the Markov model is obtained by structuring the transition probability matrix on the basis of changes of state only, so that no transitions from a given state to itself are permitted. In this arrangement the sequence of observations listed earlier becomes simply ABCBABC..., thus recording only the sequence of rock types in the stratigraphic section.

The result of structuring the data this way is to reduce the diagonal elements in the transition matrix to zero. Because this model behaves very differently from the Markov chain with transition matrix $P = [p_{ij}]$, Krumbein and Dacey use the symbol r_{ij} for the transition probabilities, with r_{ii} identically zero. Thus the earlier matrix becomes:

$$\text{From State } i \quad \begin{matrix} \\ A \\ B \\ C \end{matrix} \overset{\text{To State } j}{\begin{matrix} A & B & C \\ \end{matrix}} \begin{bmatrix} 0 & r_{AB} & r_{AC} \\ r_{BA} & 0 & r_{BC} \\ r_{CA} & r_{CB} & 0 \end{bmatrix}$$

The r's here are probabilities, not correlation coefficients.

This is called an embedded Markov chain with transition matrix $R = [r_{ij}]$. It is obtained from the P matrix by the relation $r_{ij} = p_{ij}/(1 - p_{ii})$ for all $j \neq i$. Each p_{ii} in the diagonal is then changed to zero. The embedded chain does not specify a waiting time, which means that any frequency distribution can be used as follows: The R matrix is used to get the succession of states, and for each occurrence a random observation is drawn from the frequency distribution of elements in the corresponding state.

Carr et al. (1966) used such a matrix in a study of a Chester (upper Mississippian) section, using lognormal thickness distributions for each component. Although the embedded chain is independent of the geometric waiting time, the matrix nevertheless should be tested for the Markov property. An interesting question to be raised is what a simulated section would look like if the R matrix is used directly for simulation, without assigning random thicknesses.

It should be mentioned here that probability trees and fixed probability vectors for embedded Markov chains with matrix R do not have the same interpretation as in Markov chains with matrix P. When the embedded matrix (with its zero diagonal elements) is raised to its equilibrium power, the fixed probabilities x 100 give the percentage of the overall <u>number</u> of times that a given lithologic type occurs. This is because the R matrix gives no information about thicknesses.

In their study of the Chester section, Carr et al. noted that occasionally a given rock type was immediately followed by a variant of the same kind of rock, as when a thickbedded limestone is overlain directly by a thinbedded limestone. To cope with these situations, entries were put into the diagonal of the R matrix, to represent the probability of such occurrences among the lithologic units in their section. This variant was called a <u>multistory transition</u> matrix, and it was used in simulation as before, drawing random thicknesses as needed.

The introduction of any finite element into the diagonal of an embedded matrix immediately introduces a geometric waiting time into that state of the system, and random thicknesses drawn from a lognormal distribution are no longer strictly appropriate. This can easily be seen by simulation experiments, which automatically yield geometric thickness distributions with parameter $(1 - r_{ii})$. Potter and Blakely later (1967) used this same kind of matrix to simulate a fluviatile sandstone section with several varieties of sand bedding.

The problems raised by adoption of either the simple Markov chain with matrix $P = [p_{ij}]$ or the embedded chain with matrix $R = [r_{ij}]$ mainly concern the "true" distribution of bed thicknesses and lithologic-unit thicknesses in stratigraphic sections. What has been suggested is a re-examination of the operational definitions by which these distributions are obtained, inasmuch as the critical point involved is the relative frequencies of very thin beds (Krumbein, 1972). If current operational definitions are insensitive to very thin beds or units, an observed distribution could appear to be lognormal rather than geometric, or more properly exponential in the continuous case, inasmuch as the continuous equivalent of the geometric distribution is simply $f(t) = \beta e^{\beta t}$, where β is the parameter, related to the transition probabilities in the P matrix.

4.7 EXTENSIONS

Two aspects of Markovian analysis in stratigraphy will probably become more important as time goes on. One of these, touched upon earlier, is the use of transition rates rather than transition probabilities. This involves moving the model from discrete to continuous time, but allowing the states to remain discrete. The subject is explored in Krumbein (1968a, b) in terms of a lateral-shift model that can be applied to transgressive-regressive movements of a strand line. This can actually be done with transition probabilities by redefining the states of the system as the successive positions through time of the strand line, as monitored in terms of strand-line deposits. In this approach one may start to analyze processes rather than responses, though observational data may be less readily obtained. Part of the difficulty involves exact relations between time and thickness of stratigraphic units, if the outcome of a continuous-time process is to be expressed in rock thicknesses as continuous distributions.

A second promising avenue for further research is to examine stratigraphic sections at the individual-bed level rather than that of the rock units themselves. In this approach the transition matrix is based on successions of individual beds, and the transition probabilities express the liklihood that a given kind of bed (say of shale) persists through a number of successive clock ticks, or whether the state changes to another kind of bed, say of limestone. This model involves three frequency distributions, one representing the distribution of the number of beds per lithologic unit, the second representing the thickness distributions of the individual beds, and the third concerns the thickness distributions of the lithologic units, whose individual thicknesses represent the sum of all the bed thicknesses in the unit.

Dacey and Krumbein (1970) have looked into this problem, and the most interesting part of the study was the demonstration that if the number of beds in lithologic units of the i-th lithology are distributed geometrically, and if the thickness distribution of beds of the i-th kind of lithology in the lithologic unit is also geometric (with the same or different parameters), then the thickness distribution of the lithologic units of the i-th kind of lithology will be distributed geometrically with a parameter predictable from the parameters of the number of beds distribution and the bed thickness distribution.

Implementation of the lateral-shift model and of the bed-transition model is hindered by lack of observational data in the literature. The lateral-shift model requires identification of cross-cutting relations between time lines and rock lines over short distrances, and the bedding model is presently hindered by lack of truly discriminatory ways of distinguishing between thickness samples drawn from logarithmic and exponential distributions. Some examples of both of these kinds of distributions tend to plot as relatively straight lines on log-probability paper, and even chi-square tests may not be fully discriminatory.

Despite the difficulties that beset advanced analytical applications of Markov models, in contrast to their largely descriptive present use, there are several relatively straightforward criteria that can be helpful in choosing Markovian models as against independent-events models for analyzing earth-science data. Basically these depend upon two considerations: presence or absence of a Markov property; and presence or absence of geometric distributions in the data of concern. Four combinations can be distinguished in stratigraphic analysis:

1. The observed data have a first-order Markov dependency (i.e., event t_{n+1} is controlled by the state of the system at event t_n) in the succession of lithologies, and they each have a geometric distribution of lithologic-unit thicknesses.
2. The observed data have a first-order Markov dependency in the succession of lithologies, but they do not have a geometric distribution of lithologic-unit thicknesses.

3. The observed data do not have a first-order Markov dependency in the succession of lithologies, but they do have a geometric distribution of lithologic-unit thicknesses.
4. The observed data have neither a first-order Markov dependency in the succession of lithologies nor do they have a geometric distribution of lithologic-unit thicknesses.

If combination 1 obtains, then all operations and interpretations that apply to discrete-state discrete-time first-order Markov chains with transition matrix P are appropriate. If combination 2 obtains, then the appropriate model is the embedded Markov chain with transition matrix R.

Where combination 3 obtains, the appropriate model is an independent-events model of the kind shown in equation (4.2.2).

Combination 4, having neither the Markov property nor the geometric distribution, is outside the limits of this discussion. In the present context, however, this last case could be called a "degenerate Markov chain," just as, in a sense, Anderson and Goodman's (1957) test is that of H_0 equals a "zero-order chain" as against H_a equals a first-order chain.

REFERENCES

Anderson, T. W., and Goodman, L. A., 1957, Statistical inference about Markov chains: Anals Math. Statistics, v. 28, p. 89-110.

Carr, D. D., and others, 1966, Stratigraphic sections, bedding sequences, and random processes: Science, v. 154, no. 3753, p. 1162-64.

Dacey, M. F., and Krumbein, W. C., 1970, Markovian models in stratigraphic analysis: Math. Geol., v. 2, p. 175-191.

Harbaugh, J. W., and Bonham-Carter, G., 1970, Computer simulation in geology: New York, John Wiley & Sons, 575 p.

James, W. R., and Krumbein, W. C., 1969, Frequency distributions of stream link lengths: Jour. Geology, v. 77, p. 544-565.

Krumbein, W. C., 1967, FORTRAN IV computer programs for Markov chain experiments in geology: Kansas Geol. Survey Computer Contr. 13, 38 p.

Krumbein, W. C., 1968a, Computer simulation of transgressive and regressive deposits with a discrete-state, continuous-time Markov model, in computer applications in the earth sciences: Colloquium on simulation, D. F. Merriam, ed.: Kansas Geol. Survey Computer Contr. 22, p. 11-18.

Krumbein, W. C., 1968b, FORTRAN IV computer program for simulation of transgression and regression with continuous-time Markov models: Kansas Geol. Survey Computer Contr. 26, 38 p.

Krumbein, W. C., 1972, Probabilistic models and the quantification process in geology: Geol. Soc. Amer. Spec. Paper 146, p. 1-10.

Krumbein, W. C., and Dacey, M. F., 1969, Markov chains and embedded Markov chains in geology: Math. Geol., v. 1, p. 79-96.

Pattison, A., 1965, Synthesis of hourly rainfall data: Water Resources Research, v. 1, p. 489-498.

Schwarzacher, W., 1967, Some experiments to simulate the Pennsylvanian rock sequence of Kansas: Kans. Geol. Survey Computer Contr. No. 18, p. 5-14.

BIBLIOGRAPHY

Adelman, I. G., 1958, A stochastic analysis of the size distribution of firms: Jour. Amer. Stat. Assoc., v. 53, p. 893-904. (Example of discrete states with unequal class intervals.)

Agterberg, F. P., 1966, The use of multivariate Markov schemes in geology: Jour. Geology, v. 74, p. 764-785.

Agterberg, F. P., 1966, Markov schemes for multivariate well data: Min. Ind. Experiment Sta., Pennsylvania State Univ. Spec. Publ. 2-65, p. Y1-Y18. (Theory and application of first-order Markov process to study of chemical elements in a reef.)

Allegre, C., 1964, Vers une logique mathematique des series sedimentaires: Bull. Soc. Geol. France, v. 6, p. 214-218.

Amorocho, J., and Hart, W. E., 1964, Critique of current methods in hydrologic systems investigation: Trans. Amer. Geophys. Union, v. 45, p. 307-321. [First-order and higher-order Markov chains (p. 318).]

Bartlett, M. S., 1960, An introduction to stochastic processes with special reference to methods and applications: Cambridge, The University Press, 312 p.

Billingsley, P., 1961, Statistical methods in Markov chains: Ann. Math. Stat., v. 32, p. 12-40.

Clark, W. A. V., 1964, Markov chain analysis in geography: an application to the movement of rental housing areas: Ann. Assoc. Am. Geog., v. 55, p. 351-359. (Study of rentals in several cities for three 10-year intervals.)

Coleman, J. S., 1964, Introduction to mathematical sociology: Illinois, Free Press, Glencoe, 554 p.

Doob, J. L., 1953, Stochastic processes: New York, John Wiley & Sons, Inc., 154 p.

Feller, W., 1968, An introduction to probability theory and its applications (3rd ed.): New York, John Wiley & Sons, 509 p.

Fenner, P., (Editor), 1969, Models of geologic processes: AGI/CEGS Short Course, Philadelphia, November, 1969. Available through American Geological Institute, Washington, D.C.

Gingerich, P. D., 1969, Markov analysis of cyclic alluvial sediments: Jour. Sed. Pet., v. 39, no. 1, p. 330-332.

Graf, D. L., Blyth, C. R., and Stemmler, R. S., 1967, One-dimensional disorder in carbonates: Illinois Geol. Survey Circ. 408, 61 p. (First-order Markov model applied to crystallographic defects in carbonate crystals.)

Griffiths, J. C., 1966, Future trends in geomathematics: Pennsylvania State Univ., Mineral Industries, v. 35, p. 1-8.

Harbaugh, J. W., 1966, Mathematical simulation of marine sedimentation with IBM 7090/7094 computers: Kansas Geol. Survey Computer Contr. 1, 52 p.

Harbaugh, J. W., and Wahlstedt, W. J., 1967, FORTRAN IV program for mathematical simulation of marine sedimentation with IBM 7040 or 7094 computers: Kansas Geol. Survey Computer Contr. 9, 40 p.

Heller, R. A., and Shinozuka, M., 1966, Development of randomized load sequences with transition probabilities based on a Markov process: Technometrics, v. 8, p. 107-114.

Karlin, S., 1966, A first course in stochastic processes: New York, Academic Press, 502 p.

Kemeny, J. G., and Snell, J. L., 1960, Finite Markov chains: Princeton New Jersey, Van Nostrand Co., Inc., 210 p.

Krumbein, W. C., and Graybill, F. A., 1965, An introduction to statistical models in geology: New York, McGraw-Hill Book Co., 475 p.

Krumbein, W. C., and Scherer, W., 1970, Structuring observational data for Markov and semi-Markov models in geology: Tech. Rept. No. 15, ONR Task 389-150. National Clearinghouse No. AD 716794.

Leopold, L. B., Wolman, M. G., and Miller, J. P., 1964, Fluvial processes in geomorphology: San Francisco, Freeman and Co., 522 p.

Loucks, D. P., and Lynn, W. R., 1966, Probabilistic models for predicting stream quality: Water Resources Research, v. 2, p. 593-605.

Lumsden, D. N., 1971, Facies and bed thickness distributions of limestones: Jour. Sed. Pet., v. 41, p. 593-598.

Matalas, N. C., 1967, Some distribution problems in time series simulation: Computer Contribution 18, Kansas Geol. Survey, p. 37-40.

Merriam, D. F., and Cocke, N. C., eds., 1968, Computer applications in the earth sciences: Colloquium on simulation: Kansas Geol. Survey Computer Contr. 22, 58 p.

Potter, P. E., and Blakely, R. E., 1967, Generation of a synthetic vertical profile of a fluvial sandstone body. J. Soc. Petrol. Eng., v. 6, p. 243-251.

Potter, P. E., and Blakely, R. F., 1968, Random processes and lithologic transitions: Jour. Geology, v. 76, p. 154-170.

Rogers, A., 1966, A Markovian policy model of interregional migration: Regional Sci. Assoc. Papers, v. 17, p. 205-224. (Interregional migration under controlled and uncontrolled political conditions.)

Scheidegger, A. E., and Langbein, W. B., 1966, Probability concepts in geomorphology: U.S. Geol. Survey Prof. Paper 500-C, p. C1-C14. (Markov processes continuous in time and space for slope development.)

Schwarzacher, W., 1964, An application of statistical time-series analysis of a limestone-shale sequence: J. Geol., v. 72, p. 195-213.

Schwarzacher, W., 1968, Experiments with variable sedimentation rates, in computer applications in the earth sciences: Colloquium on simulation, D. F. Merriam, ed.: Kansas Geol. Survey Computer Contr. 22, p. 19-21.

Schwarzacher, W., 1972, The semi-Markov process as a general sedimentation model: Mathematical models in sedimentology, edited by D. F. Merriam: New York, Plenum Press, p. 247-268.

Shreve, R. L., 1966, Statistical law of stream numbers: Jour. Geology, v. 74, p. 17-37.

Shreve, R. L., 1967, Infinite topologically random channel networks: Jour. Geology, v. 75, p. 178-186.

Shreve, R. L., 1969, Stream lengths and basin areas in topologically random channel networks: J. Geol., v. 77, p. 397-414.

Smart, J. S., 1968, Statistical properties of stream lengths: Water Resources Research, v. 4, p. 1001-1014.

Smart, J. S., 1969, Topological properties of channel networks: Geol. Soc. America Bull., v. 80, p. 1757-1774.

Vistelius, A. B., 1949, On the question of the mechanism of the formation of strata: Doklady Akademii, Nauk SSSR, v. 65, p. 191-194.

Vistelius, A. B., and Feigel'son, T. S., 1965, On the theory of bed formation: Doklady Akademii, Nauk SSSR, v. 164, p. 158-160.

Vistelius, A. B., and Faas, A. V., 1965, On the character of the alternation of strata in certain sedimentary rock masses: Doklady Akademii, Nauk SSSR, v. 164, p. 629-632.

Vistelius, A. B., 1966, Genesis of the Mt. Belaya granodiorite, Kamchatka (an experiment in stochastic modeling): Doklady Akademii, Nauk SSSR, v. 167, p. 1115-1118. (Application of a Markov chain in the study of the sequence of mineral grains in a thin section.)

Watson, R. A., 1969, Explanation and prediction in geology: Jour. Geology, v. 77, p. 488-494.

Wickman, F. E., 1966, Repose period patterns of volcanoes; V. General discussion and a tentative stochastic model: Arkiv Mineralogi Geologi, v. 4, p. 351-367.

Zeller, E. J., 1964, Cycles and psychology, in Symposium on Cyclic Sedimentation: Kansas Geol. Survey Bull. 169, v. 2, p. 631-636.

Chapter 5

A Priori and Experimental Approximation of Simple Ratio Correlations

Felix Chayes

5.1 RATIO CORRELATIONS

Most measurements are of quantities that are in some sense ratios, but this requires no special consideration in correlation analysis or in studies of interdependence if the denominators of the ratios being compared are constants. That elevation is measured in units initially defined as some fraction of the distance from the equator to the pole and specific gravity as some multiple of the weight of an equivalent volume of water at a particular temperature and pressure, for instance, need not concern the petrologist seeking to characterize the relation between the elevations of a set of samples in a sill and their specific gravities. Indeed, without such scaling parameters it is difficult to see how questions of this kind could be answered, or even asked.

When the scaling parameters are themselves variables, however, as is often the case in geochemistry, the situation is very different. Relations between ratios may then be very different from those between the numerators and denominators - the "absolute" variables or "terms" - of the ratios. In particular, as was noted long ago by Pearson (1896), even though the terms are uncorrelated, there may nevertheless be correlation, and sometimes very strong correlation, between pairs of ratios formed from them.

To characterize the general relationship, we first express the ratios $Y_i = X_1/X_2$, $Y_j = X_3/X_4$ as first-order approximations of the true means, variances, and covariances of the X's, where each "observation" vector, $\mathbf{X} = [X_1,X_2,X_3,X_4]$ is drawn simply at random from a parent population characterized by means and variances μ_m, σ_m^2 for m = 1,4 and covariances σ_{mn} for $m \neq n$.

Each observed value of $X_m = \mu_m + \delta_m$, so that

$$Y_i = \frac{\mu_1 + \delta_1}{\mu_2 + \delta_2} \text{ and } Y_j = \frac{\mu_3 + \delta_3}{\mu_4 + \delta_4} \tag{5.1.1}$$

It is readily shown (see, for instance, Chayes, 1971) that to first-order approximation

$$Y_i \simeq \frac{1}{\mu_2}(\mu_1 + \delta_1)\left(1 - \frac{\delta_2}{\mu_2}\right) = \frac{\mu_1}{\mu_2} + \frac{1}{\mu_2^2}(\mu_2\delta_1 - \mu_1\delta_2) \tag{5.1.2}$$

and, similarly,

$$Y_j \simeq \frac{\mu_3}{\mu_4} + \frac{1}{\mu_4^2}(\mu_4\delta_3 - \mu_3\delta_4) \tag{5.1.3}$$

Then, taking expectations on both sides of (5.1.2) and (5.1.3),

$$E(Y_i) \simeq \frac{\mu_1}{\mu_2} \text{ and } E(Y_j) \simeq \frac{\mu_3}{\mu_4} \tag{5.1.4}$$

so that, using (5.1.2) and the left half of (5.1.4),

$$\Delta_i = Y_i - E(Y_i) \simeq \frac{1}{\mu_2^2}(\mu_2\delta_1 - \mu_1\delta_2) \tag{5.1.5}$$

and, from (5.1.3) and the right half of (5.1.4),

$$\Delta_j = Y_j - E(Y_j) \simeq \frac{1}{\mu_4^2}(\mu_4\delta_3 - \mu_3\delta_4) \tag{5.1.6}$$

Further, multiplying (5.1.5) by (5.1.6)

$$\Delta_i\Delta_j \simeq \frac{1}{\mu_2^2\mu_4^2}(\mu_2\mu_4\delta_1\delta_3 - \mu_2\mu_3\delta_1\delta_4 - \mu_1\mu_4\delta_2\delta_3 + \mu_1\mu_3\delta_2\delta_4) \tag{5.1.7}$$

To find the parent correlation, ρ_{ij}, between Y_i and Y_j, we require the expectations of Δ_i^2, Δ_j^2, $\Delta_i\Delta_j$, viz.,

$$\mathrm{Var}(Y_i) = E(\Delta_i^2) \simeq E\left[\frac{1}{\mu_2^4}(\mu_2^2\delta_1^2 + \mu_1^2\delta_2^2 - 2\mu_1\mu_2\delta_1\delta_2)\right]$$

$$\simeq \frac{1}{\mu_2^4}(\mu_2^2\sigma_1^2 + \mu_1^2\sigma_2^2 - 2\mu_1\mu_2\sigma_{12}) \qquad (5.1.8)$$

$$\mathrm{Var}(Y_j) = E(\Delta_j^2) \simeq \frac{1}{\mu_4^4}(\mu_4^2\sigma_3^2 + \mu_3^2\sigma_4^2 - 2\mu_3\mu_4\sigma_{34}) \qquad (5.1.9)$$

and

$$\mathrm{Cov}(Y_i, Y_j) = E(\Delta_i\Delta_j)$$

$$\simeq \frac{1}{\mu_2^2\mu_4^2}(\mu_2\mu_4\sigma_{13} - \mu_2\mu_3\sigma_{14} - \mu_1\mu_4\sigma_{23} + \mu_1\mu_3\sigma_{24}) \qquad (5.1.10)$$

Thus, finally,

$$\rho_{ij} = \frac{\mathrm{Cov}(Y_i, Y_j)}{\sqrt{\mathrm{Var}(Y_i) \cdot \mathrm{Var}(Y_j)}}$$

$$\simeq \frac{\mu_2\mu_4\sigma_{13} - \mu_2\mu_3\sigma_{14} - \mu_1\mu_4\sigma_{23} + \mu_1\mu_3\sigma_{24}}{\sqrt{(\mu_2^2\sigma_1^2 + \mu_1^2\sigma_2^2 - 2\mu_1\mu_2\sigma_{12})\,(\mu_4^2\sigma_3^2 + \mu_3^2\sigma_4^2 - 2\mu_3\mu_4\sigma_{34})}}$$

(5.1.11)

Now by definition $\sigma_{mn} = \sigma_m\sigma_n\rho_{mn}$, so that division of the numerator and denominator of (5.1.11) by $(\mu_1\mu_2\mu_3\mu_4)$ leads at once to the commonly found form [see, for instance, equation (2.1) of Chayes, 1971, in which the signs of the correlation terms in the denominator are wrong]

$$\rho_{ij} \simeq \frac{C_1C_3\rho_{13} - C_1C_4\rho_{14} - C_2C_3\rho_{23} + C_2C_4\rho_{24}}{\sqrt{(C_1^2 + C_2^2 - 2C_1C_2\rho_{12})\,(C_3^2 + C_4^2 - 2C_3C_4\rho_{34})}} \qquad (5.1.12)$$

where $C_m = \sigma_m/\mu_m$ is Pearson's coefficient of variation and ρ_{mn} is his coefficient of correlation.

From (5.1.12) it is evident, as is intuitively obvious, that if all terms of a pair of ratios are different and uncorrelated, the ratios themselves will also be uncorrelated. If, however, two ratios have a common denominator they will be correlated even though their numerators are uncorrelated with each other and with the denominator. This at first sight paradoxical result - called "spurious" correlation by Pearson - can be reached by introducing into (5.1.11) or (5.1.12) the constraints that $\mu_2 = \mu_4$, $\sigma_2 = \sigma_4$, and $\rho_{mn} = 0$ for all $m \neq n$. But working it out *ab initio* is just as simple and provides useful drill for the novice.

If $Y_i = X_1/X_2$ and $Y_k = X_3/X_2$, then of course Δ_i is exactly as in (5.1.5) and

$$\Delta_k \simeq \frac{1}{\mu_2^2}(\mu_2\delta_3 - \mu_3\delta_2) \tag{5.1.13}$$

so that

$$\Delta_i\Delta_k = \frac{1}{\mu_2^4}(\mu_2\delta_1 - \mu_1\delta_2)(\mu_2\delta_3 - \mu_3\delta_2) \tag{5.1.14}$$

If the X's are uncorrelated, the expectations of all cross product terms in the δ's vanish, and

$$\mathrm{Var}(Y_i) \simeq \frac{1}{\mu_2^4}(\mu_2^2\sigma_1^2 + \mu_1^2\sigma_2^2) \tag{5.1.15}$$

$$\mathrm{Var}(Y_k) \simeq \frac{1}{\mu_2^4}(\mu_2^2\sigma_3^2 + \mu_3^2\sigma_2^2) \tag{5.1.16}$$

$$\mathrm{Cov}(Y_i, Y_k) = E(\Delta_i\Delta_k) \simeq \frac{\mu_1\mu_3}{\mu_2^4}\sigma_2^2 \tag{5.1.17}$$

Thus,

$$\rho_{ik} = \frac{\mathrm{Cov}(Y_i, Y_k)}{\sqrt{\mathrm{Var}(Y_i) \cdot \mathrm{Var}(Y_k)}}$$

$$\simeq \frac{\mu_1\mu_3\sigma_2^2}{\sqrt{(\mu_2^2\sigma_1^2 + \mu_1^2\sigma_2^2)(\mu_2^2\sigma_3^2 + \mu_3^2\sigma_2^2)}} \tag{5.1.18}$$

which approximates the correlation between two ratios with common denominator as a function of the means and variances of the numerators and denominator, a result again easily restated in terms of the coefficients of correlation and variation.*

Thus, ratios with common denominator will tend to be positively correlated if their numerators are uncorrelated with each other and with their denominator. Indeed, the correlation generated in this fashion may be far from trivial. If, for instance, the coefficients of variation of the terms are equal, the correlation between the ratios is ~0.5, and if the coefficient of variation of the denominator is larger than the (equal) coefficients of variation of the numerators, the correlation between the ratios will be greater than 0.5; if it is twice as large, something not at all unlikely in geochemistry, the correlation of the ratios will be ~0.8.

The other simple ratio correlations - those between a ratio and its numerator or denominator, between ratios with common numerator, and between ratios the numerator of one of which is the denominator of the other - can of course be approximated in analogous fashion. All save the last are common in geochemical work (for a review, see Chayes, 1949) and the interested reader will find it useful to carry through the computations for the case of zero covariance between the X's, comparing his results with those shown in Table 2.1 of Chayes (1971).

The approximations used here and in all the work so far cited are of first order only, and can be expected to yield reliable results only if terms of second and higher order in (δ_2/μ_2) are small enough to ignore. In much geochemical work this is not so; indeed, in this field we often seem to use a particular variable as denominator precisely because its relative variance is large, so that higher powers of (δ_2/μ_2) will often not be negligible. Although the work up to this point

*Division of the numerator and denominator of the right side of (5.1.18) by $(\mu_1^2\mu_2^2\mu_3^2)$ leads at once to

$$\rho_{ik} \simeq \frac{C_2^2}{\sqrt{(C_1^2 + C_2^2)\ (C_2^2 + C_3^2)}}$$

shows pretty clearly that correlation generated by the process of ratio formation may be far too strong to ignore, in many practical cases, alas, it will not lead to useful approximations of that correlation.

Higher-order approximations for means, variances, and covariances are available (Tukey), and an analytical formulation using them would have the usual advantage of providing, in principle at least, a general solution, something of considerable aesthetic and scientific appeal. But a general solution is not indispensable if one can obtain a satisfactory solution for any specific problem that may arise. That is what one ought to be able to accomplish by simulation experimentation, and the remainder of this chapter describes the structure and use of a computer program, RTCRSM2 (Ratio Correlation Simulation, version 2), which is an attempt to exploit this possibility.

5.2 RTCRSM2

In using RTCRSM2, the investigator assigns:

1. Appropriate parent means and variances to the four pseudorandom variables A,B,C,D to be used as numerators and denominators of the ratios
2. The number of items (sample size) per simulation, and the number of simulations per experiment

He must also initialize the random number generator, either by providing it with a starting residue or instructing it to use a stored one.

Given this information, the program generates a set of four random numbers from a parent population uniformly distributed in the range (0,1), transforms these to normal deviates with zero mean and unit variance, and adjusts each with its assigned mean and variance to produce the current set of "observed" values of A, B, C, and D. From these all possible simple ratios are formed, and the elements of this vector of terms and ratios, together with their squares and cross products, are then stored in cumulators. The process is repeated until the requested number of items - i.e., sets of "observed values" of A, B, C, D - has been supplied and processed.

The covariance matrix is then computed from the cumulated sums, sums of squares, and sums of cross products, its diagonal elements are converted to standard deviations, its off-diagonal elements to correlations, and the requested results printed out. Since the objective is to approximate as closely as possible the value of an unknown parent correlation, the number of items per simulation should be as large as the computing budget will permit. But the cumulating procedure used in the program, selected because it economizes on core requirement and places no upper limit on the number of items per sample, leads to large rounding error. Double precision is essential for simulations containing more than a few hundred items; on the Univac 1108 it seems to satisfactorily control rounding error even for very large simulations.

In this kind of work it is easy to bury oneself in numbers. RTCRSM2 is designed as a specific problem solver, and its printout may be restricted to those particular ratio correlations of immediate interest. In fact, unless the user specifies the type(s) of correlation(s) to be printed, the output will consist only of an error message reminding him that he should have done so. Loading instructions are provided in lines 32 to 76 of the accompanying program listing.

5.3 UNNO AND RANEX

The random number generator referred to in RTCRSM2 as UNNO is designed to generate uniformly distributed numbers in the range (0,1); these are normalized in the main program. UNNO, coded in Fortran by L. Finger, appears to work admirably on the Univac 1108 used for the calculations reported below. Whether it performs satisfactorily on any specific computer can be determined experimentally and such experimentation should certainly precede routine operation of RTCRSM2, for unless the random number generator it uses is demonstrably sound, the results yielded by RTCRSM2 are uninterpretable. Program RANEX is designed to provide information on this matter; the rather extensive battery of tests performed by it is described in lines 2 to 9, and loading instructions are given in lines 19 to 34 of the accompanying listing. If UNNO performs unsatisfactorily or the user prefers another generator, the subroutine calls in RTCRSM2 will require modification: these are contained in lines 231, 249, and 250. (With analogous modification of cards 98, 110, and 111, incidentally, RANEX may be used to test the output of any subroutine designed to generate uniformly distributed pseudorandom numbers in the range 0,1.)

5.4 COMMENTS ON USAGE

Unless instructions to the contrary are provided at operation time, the variables A, B, C, D of RTCRSM2 are drawn from theoretically uncorrelated parents; in reasonably large samples, the correlations between them should be negligibly small. The actual sample correlations can be printed out, however, so it is always possible to see how nearly this goal has been attained in a particular simulation, and it is probably wise to do so. But ratio correlations can be approximated by simulation whatever the correlations between their terms, and RTCRSM2 provides limited facility for relaxing the restriction that the terms are uncorrelated. Specifically, the user may assign correlation(s) of arbitrary size and sign between any pair or any two mutually exclusive pairs of variables A, B, C, and D. (When this option is exercised, the sample correlations between variables A, B, C, and D should always be printed out.)

Preliminary experimentation with RTCRSM2 confirms an earlier suggestion (Chayes, 1971) that for ratios formed from uncorrelated terms homogeneous in C, the linear approximations of ρ are very good if $C < 0.1$ and still fairly good if $C < 0.15$. For $0.15 < C < 0.35$ the simulated ratio correlations differ widely both from the linear approximations and from zero, so that experimental determination or higher-order approximation of null values against which to test observed ratio correlations will be essential in this range. Simulation experiments suggest that with further increase of C, Pearson's "spurious" correlation - that between ratios with common denominator - averages about 0.5 with very large variance, while the other simple ratio correlations rapidly approach zero with small variance.

These rather unexpected results will be described more fully when the work is completed and are mentioned here only for the sake of the perspective they provide. Compositional variables are necessarily nonnegative and in the absence of pronounced skew their coefficients of variation must be fairly small. For uniformly distributed variables - in which there is no central tendency - it is easily shown that $C = \sqrt{1/3}$ or 0.557; present indications are that it would be wise to avoid correlations between ratios of such variables with common denominators but safe enough to test observed values of other simple ratio correlations formed from them against a null value of zero. In binomial or multinomial variables, on the other hand, $C = \sqrt{(1 - p)/np}$; here the first-order approximations of ratio correlations should be adequate if $0.05 < p < 0.95$ and $n > 400$, and if $p > 0.1$ this should be true even for n as small as 100. But in work involving sampling variation, as opposed to mere counting variance, of major constituents, the relevant values of C will nearly always be considerably larger than for binomially distributed variables and, since there is usually a fairly strong central tendency, considerably smaller than for rectangularly distributed ones. It seems likely, then, that in dealing with ratios formed of major constituents, whether expressed normatively, modally, or as oxides, there may be frequent need for experimental determination of the appropriate null value of ρ along the lines suggested here, for it will often happen that the coefficients of variation of such variables are large enough to make the first-order approximation of ρ unsatisfactory, but small enough so that the assumption that $\rho = 0$ is unrealistic.

REFERENCES

Chayes, F., 1949, On correlation in petrography: J. Geol., v. 57, p. 239-254.

Chayes, F., 1971, Ratio correlation: Chicago, University of Chicago Press.

Pearson, K., 1896-1897, On a form of spurious correlation which may arise when indices are used in the measurements of organs: Proc. Roy. Soc. (London), v. 60, p. 489-502.

Tukey, J. W., undated, The propagation of errors, fluctuations and tolerances: Unpublished Technical Reports No. 10, 11, 12, Princeton University.

APPENDIX 1

```
C                 PROGRAM RTCRSM2                                       RTCR  10
C FINDS CORRELATIONS BETWEEN ALL RATIOS FORMED FROM 4 TERMS (A,B,C,D)   RTCR  20
C WITH ASSIGNED MEANS,STANDARD DEVIATIONS AND COMMON ELEMENTS, BY MONTE RTCR  30
C CARLO SIMULATION.  FOUR UNIFORMLY DISTRIBUTED PSEUDORANDOM NUMBERS    RTCR  40
C ARE GENERATED IN THE RANGE (0,1), NORMALIZED BY THE 'DIRECT'          RTCR  50
C PROCEDURE OF ZELEN AND SEVERO,ADJUSTED FOR ASSIGNED MEANS, STANDARD   RTCR  60
C DEVIATIONS AND COMMON ELEMENTS, AND STORED IN THE FIRST FOUR ELEMENTS RTCR  70
C OF VECTOR Y (A=Y(1), B=Y(2),C=Y(3),D=Y(4)). THE REMAINING 12 ELEMENTS RTCR  80
C OF Y ARE LOADED WITH BINARY RATIOS OF THE FIRST 4, IN THE ORDER SHOWN RTCR  90
C BELOW IN DATA BLOCK NAM. AS EACH Y VECTOR IS COMPLETED, THE SUMS, SUMSRTCR 100
C OF SQUARES AND SUMS OF CROSS-PRODUCTS OF ITS ELEMENTS ARE CUMULATED.  RTCR 110
C THE PROCEDURE IS ITERATED NR TIMES TO GENERATE THE SAMPLE ARRAY.  ON  RTCR 120
C COMPLETION OF THE NRTH CYCLE THE COVARIANCE MATRIX IS GENERATED       RTCR 130
C FROM THE CUMULATED SUMS OF SQUARES AND CROSS-PRODUCTS, AND THE DESIREDRTCR 140
C CORRELATIONS ARE EXTRACTED AND PRINTED.                               RTCR 150
C                                                                       RTCR 160
C   IF NO CORRELATIONS ARE SPECIFED ON INPUT, VARIABLES A,B,C,D WILL BE RTCR 170
C UNCORRELATED, TO LIMITS OF EXPERIMENTAL ERROR. IF USER SPECIFIES      RTCR 180
C DESIRED CORRELATIONS (POSITIVE OR NEGATIVE) BETWEEN ANY PAIR OR ANY 2 RTCR 190
C MUTUALLY EXCLUSIVE PAIRS OF VARIABLES A,B,C,D, THE REQUIRED COMMON EL-RTCR 200
C EMENT STANDARD DEVIATIONS ARE COMPUTED FROM AN ALGORITHM BASED ON     RTCR 210
C EQ. (3.15), P.26, OF RATIO CORRELATION (CHAYES,1971). EXECUTION TERM- RTCR 220
C INATES IN ERROR IF THE SAME VARIABLE APPEARS IN BOTH CORRELATIONS.    RTCR 230
C                                                                       RTCR 240
C RANDOM NUMBERS ARE GENERATED BY SUBROUTINE UNNO(I,F), WHERE I IS THE  RTCR 250
C INTEGER SEED AND F IS THE FLOATED NUMBER IN RANGE (0,1). THE FORTRAN  RTCR 260
C GENERATOR USED WITH THIS VERSION OF RANEX WAS CODED BY L. FINGER.     RTCR 270
C                                                                       RTCR 280
C PROGRAM WRITTEN BY F. CHAYES FOR NSF INSTITUTE ON GEOSTATISTICS,      RTCR 290
C                        CHICAGO CIRCLE, 1972                           RTCR 300
C                                                                       RTCR 310
```

```
C    ********************************************************************* RTCR 320
C    *                                                                   * RTCR 330
C    *                 CARD INPUT TO PROGRAM RTCRSM2                     * RTCR 340
C    *                                                                   * RTCR 350
C    *  COMMAND CARD                              (I5,I4,11I1,I15)      * RTCR 360
C    *    COL.   VARIABLE        DEFINITION OR FUNCTION                  * RTCR 370
C    *    1-5    NR           NUMBER OF ITEMS PER SAMPLE                 * RTCR 380
C    *    6-9    NMSP         NUMBER OF SAMPLES TO BE DRAWN              * RTCR 390
C    *   10      KMLMT         0-NOP, 1 - READ AND USE COMMON ELEMENTS   * RTCR 400
C    *      (NB - VECTOR RQST RESTRICTS PRINT TO MATERIAL REQUESTED)     * RTCR 410
C    *    11     RQST(1)       0-NOP, 1 - ALL, IN ONE BIG MATRIX         * RTCR 420
C    *    12       '' (2)      0-NOP, 1 - BETWEEN ALL TERMS (A,B,C,D)    * RTCR 430
C    *    13       '' (3)      0-NOP  1 - OF TYPE X1/X2 WITH X1          * RTCR 440
C    *    14       '' (4)      0-NOP, 1 - OF TYPE X1/X2 WITH X2          * RTCR 450
C    *    15       '' (5)      0-NOP  1 - OF TYPE X1/X2 WITH X1/X3       * RTCR 460
C    *    16       '' (6)      0-NOP  1 - OF TYPE X1/X2 WITH X3/X2       * RTCR 470
C    *    17       '' (7)      0-NOP, 1 - OF TYPE X1/X2 WITH X2/X3       * RTCR 480
C    *    18       '' (8)      0-NOP, 1 - OF TYPE X1/X2 WITH X2/X1       * RTCR 490
C    *    19       '' (9)      0-NOP, 1 - OF TYPE X1/X2 WITH X3/X4       * RTCR 500
C    *    20       '' (10)               UNASSIGNED                      * RTCR 510
C    *   21-35   QIN          STARTING SEED OF RANDOM NUMBER GENERATOR   * RTCR 520
C    *                              (QIN = K' USE K                      * RTCR 530
C    *                               QIN = 0' USE 1ST RESIDUE FROM UNNO  * RTCR 540
C    *                               QIN = -1' USE EXISTING RESIDUE)     * RTCR 550
C    *                                                                   * RTCR 560
C    *  PARAMETER CARD                                    (12F6.3)       * RTCR 570
C    *    1-6    GVAV(1)      PARENT MEAN OF A                           * RTCR 580
C    *    7-12   GVSD(1)      PARENT STANDARD DEVIATION OF A             * RTCR 590
C    *     .                     .     .                                 * RTCR 600
C    *     .                     .     .                                 * RTCR 610
C    *   43-48   GVSD(4)      PARENT STANDARD DEVIATION OF D             * RTCR 620
C    *                                                                   * RTCR 630
C    *  COMMON ELEMENT CARD     (USE ONLY IF KMLMT = 1)     (12F6.3)     * RTCR 640
C    *    1-6    RHO(1)        DESIRED CORR. BTW. VARIABLES A AND B      * RTCR 650
C    *    7-12   RHO(2)        DESIRED CORR. BTW. VARIABLES A AND C      * RTCR 660
```

```
C     *    13-18    RHO(3)         DESIRED CORR. BTW. VARIABLES A AND D      *
C     *    19-24    RHO(4)         DESIRED CORR. BTW. VARIABLES B AND C      *
C     *    25-30    RHO(5)         DESIRED CORR. BTW. VARIABLES B AND D      *
C     *    31-36    RHO(6)         DESIRED CORR. BTW. VARIABLES C AND D      *
C     *          N.B. - ANY ELEMENT OF RHO, OR ANY ONE OF THE PAIRS          *
C     *          (1,6), (2,5), (3,4) MAY BE ASSIGNED NON-ZERO VALUES.        *
C     *          PROBLEMS IN WHICH OTHER PAIRS OR MORE THAN TWO ELEMENTS     *
C     *          OF RHO ARE ASSIGNED NON-ZERO VALUES WILL BE IGNORED.        *
C     *                                                                      *
C     ************************************************************************
C
      DIMENSION AV(32), RHO(6), CMSD(6), GVAV(4), GVSD(4), NAM(16),
     *PRBF(36), R(4), SIG(32,32), SUM(32), Y(32), RQST(10),NUPG(2),
     *ORSD(4)
      DOUBLE PRECISION AV,SUM,SIG,RN,RNL,Y
      REAL NAM
      INTEGER OUT,Q,QIN,RQST
      DATA IN,OUT,NUPG,/5,6,4H(1H1,1H)/
      DATA NAM/' A ',' B ',' C ',' D ','A/D','B/D','C/D','D/C','A/C',
     *'B/C','C/B','D/B','A/B','B/A','C/A','D/A'/
C
C                                 INPUT FORMATS
    1 FORMAT (I5,I4,11I1,I15)
    3 FORMAT (12F6.3)
    5 FORMAT (6(6X,F6.3))
C                                 OUTPUT FORMATS
    2 FORMAT ('1',25X,'A. - ASSIGNED PARAMETERS FOR SIMULATIONS YIELDING
     * CORRELATIONS SHOWN IN B, BELOW'//'OPARENT VALUES OF TERMS OF RATI
     *OS'/16X,4(8X,A2)/10X,'MEAN',6X,4F10.4/10X,'STD.-DEV.',F11.4,3F10.4
     *,//I10,' ITERATIONS IN SIMULATION.  ENTRY SEED FOR RANDOM NUMBER G
     *ENERATOR IS ',I15,' IN SAMPLE NUMBER',I3,'.')
    4 FORMAT (' NO ASSIGNED COMMON ELEMENTS AMONG TERMS OF RATIOS.')
    6 FORMAT (/15X,'INITIAL PARAMETERS LISTED ABOVE WILL BE MODIFIED BY
     *COMMON ELEMENT ADJUSTMENTS TO INTRODUCE FOLLOWING CORRELATIONS-'/
     *51X,6(A2,A2,7X)/' REQUESTED CORRELATIONS',27X,6(F6.4,5X)/' REQUIRE
```

```
 *D STANDARD DEVIATIONS OF COMMON ELEMENTS',6(F8.4,3X))              RTCR1020
 8 FORMAT   (/35X'COMPARISON OF PARENT AND SAMPLE VALUES OF RATIO TERRTCR1030
 *MS A, B, C AND D.'/55X,'MEAN',12X,'STD.-DEV.'/40X,'TERM      PARENRTCR1040
 *T  SAMPLE',4X,'PARENT  SAMPLE'/(40X,A2,8X,F6.2,1X,F7.3,3X,F6.2,    RTCR1050
 *1X,F8.4/))                                                         RTCR1060
10 FORMAT (/'OMATRIX WITH CORRELATIONS OFF, STANDARD DEVIATIONS ON, DRTCR1070
 *IAGONAL -'/15X,16(A3,4X))                                          RTCR1080
12 FORMAT (9X,A3,16F7.4)                                             RTCR1090
14 FORMAT (//' CORRELATIONS BETWEEN TERMS',4X,6(2X,A3,',',A3)/33X,   RTCR1100
 *6(F6.4,3X))                                                        RTCR1110
16 FORMAT (//' CORRELATIONS BETWEEN RATIOS AND THEIR NUMERATORS -'/  RTCR1120
 *12(2X,A3,',',A3,1X)/2X,F6.4,11(4X,F6.4))                           RTCR1130
18 FORMAT (//' CORRELATIONS BETWEEN RATIOS AND THEIR DENOMINATORS -'/RTCR1140
 *12(2X,A3,',',A3,1X)/2X,F6.4,11(4X,F6.4))                           RTCR1150
20 FORMAT (//' CORRELATIONS BETWEEN RATIOS WITH COMMON NUMERATORS -'/RTCR1160
 *12(2X,A3,',',A3,1X)/2X,F6.4,11(4X,F6.4))                           RTCR1170
22 FORMAT (//' CORRELATIONS BETWEEN RATIOS WITH COMMON DENOMINATORS -RTCR1180
 *'/12(2X,A3,',',A3,1X)/2X,F6.4,11(4X,F6.4))                         RTCR1190
24 FORMAT (//' CORRELATIONS BETWEEN RATIOS IN WHICH ONE TERM,',A3,'ORRTCR1200
 *',A2,', IS THE NUMERATOR OF ONE AND THE DENOMINATOR OF THE OTHER -RTCR1210
 * '/12(2X,A3,',',A3,1X)/2X,F6.4,11(4X,F6.4))                        RTCR1220
26 FORMAT (//' CORRELATIONS BETWEEN RATIOS WHICH ARE RECIPROCALS - ',RTCR1230
 *6(2X,A3,',',A3,1X)/55X,6(F6.4,4X))                                 RTCR1240
28 FORMAT (//' USE ',I15,' AS SEED TO RANDOM NUMBER GENERATOR IN NEXTRTCR1250
 * EXPERIMENT'/1H1)                                                  RTCR1260
30 FORMAT (//25X,'B. - SIMULATION RESULTS')                          RTCR1270
32 FORMAT('1',20X,'ASK ME NO QUESTIONS AND I WILL TELL YOU NO LIES. VRTCR1280
 *ECTOR RQST IS EMPTY.  TRY AGAIN '/1H1)                             RTCR1290
34 FORMAT (//' CORRELATIONS BETWEEN RATIOS WITH NO COMMON TERMS -'/  RTCR1300
 *12(2X,A3,',',A3,1X)/2X,F6.4,11(4X,F6.4))                           RTCR1310
36 FORMAT (20X,'FAULTY DATA CARD. PROGRAM LOOKS FOR NEXT PROBLEM.'/' RTCR1320
 *1')                                                                RTCR1330
38 FORMAT  ('1KMLMT CARD SPECIFIES MORE THAN TWO OR A FAULTY PAIR OF RTCR1340
 *CORRELATIONS. READ IN NEXT PROBLEM OR QUIT.')                      RTCR1350
                                                                     RTCR1360
```

```
C READ COMMAND CARD                                                     RTCR1370
   70 READ (IN,1,ERR=73,END=300) NR,NMSP,KMLMT,RQST,QIN                RTCR1380
      IF (NMSP.EQ.0) NMSP = 1                                           RTCR1390
C                                                                       RTCR1400
C CHECK PRINT REQUEST.  IF IT IS EMPTY, SKIP TO NEXT PROBLEM.           RTCR1410
      DO 72 J = 1,10                                                    RTCR1420
      IF (RQST(J).EQ.1) GO TO 75                                        RTCR1430
   72 CONTINUE                                                          RTCR1440
      WRITE (OUT,32)                                                    RTCR1450
      GO TO 70                                                          RTCR1460
C                                                                       RTCR1470
C PRINT ERROR MESSAGE, RETURN FOR NEW COMMAND CARD                      RTCR1480
   73 WRITE (OUT,36)                                                    RTCR1490
      GO TO 70                                                          RTCR1500
C INITIALIZE RANDOM NUMBER GENERATOR IF THIS IS FIRST PASS OF EXECUTION.RTCR1510
   75 IF (QIN) 100,80,90                                                RTCR1520
   80 CALL UNNO(Q,RA)                                                   RTCR1530
      GO TO 100                                                         RTCR1540
   90 Q = QIN                                                           RTCR1550
C READ PARAMETER CARD, THEN COMMON ELEMENT CARD IF REQUIRED.            RTCR1560
  100 READ (IN,3,ERR=73) (GVAV(I),GVSD(I),I=1,4)                        RTCR1570
C       REREAD GVSD INTO ORSD SO IT CAN BE RECLAIMED AS NEEDED.         RTCR1580
      READ (0,5) ORSD                                                   RTCR1590
      NS = 0                                                            RTCR1600
      IF (KMLMT.EQ.0) GO TO 101                                         RTCR1610
C READ REQUESTED CORRELATIONS BETWEEN A,B,C,D, COMPUTE REQUIRED COMMON- RTCR1620
C ELEMENT STANDARD DEVIATIONS                                           RTCR1630
      READ (IN,3,ERR=73) RHO                                            RTCR1640
C DETERMINE WHETHER CORRELATION REQUEST IS VALID.                       RTCR1650
      KR = 0                                                            RTCR1660
      DO 1002 K = 1,6                                                   RTCR1670
      IF (RHO(K)) 1001,1002,1001                                        RTCR1680
 1001 KR = KR + 1                                                       RTCR1690
 1002 CONTINUE                                                          RTCR1700
      IF (KR.EQ.1) GO TO 1007                                           RTCR1710
```

```
      IF (KR.EQ.2) GO TO 1004
C CORRELATION REQUEST FAULTY, SKIP TO NEXT PROBLEM
 1003 WRITE (OUT,38)
      GO TO 70
C DETERMINE WHETHER A VALID PAIR OF CORRELATIONS HAS BEEN REQUESTED
 1004 KR = 0
      DO 1006 K = 1,6
      IF (RHO(K)) 1005,1006,1005
 1005 KR = KR + K
 1006 CONTINUE
      IF (KR.NE.7) GO TO 1003
C PROCESS VALID CORRELATION REQUEST
 1007 KC = 0
      DO 1010 J = 1,3
      JK = J + 1
      DO 1010 K = JK,4
      KC = KC + 1
      IF (RHO(KC)) 1008,1010,1008
 1008 PF = (GVSD(K)/GVSD(J))**2
      RSQ = RHO(KC)**2
      CMPT = (RSQ*(1.+PF)+SQRT(RSQ**2*(1.-PF)**2+4.*RSQ*PF))/(2.*(1.
     *-RSQ))
      CMSD(KC) = GVSD(J)*SQRT(CMPT)
 1010 CONTINUE
  101 NS = NS + 1
      IF (NS.GT.NMSP) GO TO 70
C TRANSFER INITIAL STANDARD DEVIATIONS FROM ORSD TO GVSD
      DO 1015 I = 1,4
 1015 GVSD(I) = ORSD(I)
C RECORD INPUT DATA
      WRITE (OUT,2) (NAM(I),I=1,4),(GVAV(J),J=1,4),(GVSD(K),K=1,4),NR,Q,
     *NS
      LINE = 9
      IF(KMLMT) 102,102,103
  102 WRITE (OUT,4)
```

```
      LINE = LINE + 1
      GO TO 105
C LOAD COMMON-TERM NAME-PAIRS IN PRBF
  103 I = 0
      DO 104 J = 1,3
      JK = J + 1
      DO 104 K = JK,4
      I = I + 1
      PRBF(I) = NAM(J)
      I = I + 1
  104 PRBF(I) = NAM(K)
      WRITE (OUT,6) (PRBF(I),I=1,12),(RHO(J),J=1,6),(CMSD(K),K=1,6)
      LINE = LINE + 5
C CLEAR CUMULATORS
  105 DO 106 J = 1,16
      SUM(J) = 0.0
      DO 106 K = J,16
  106 SIG(K,J) = 0.0
C
C GENERATE ARRAY, CUMULATING SUMS OF VARIABLES, SQUARES AND CROSS-
C                             PRODUCTS STEPWISE.
      DO 145 N = 1,NR
C GENERATE ONE SET OF VALUES FOR A,B,C,D, AND STORE IN Y(I),I=1,4.
      DO 110 IR = 1,4
  110 CALL UNNO(Q,R(IR))
C
C NORMALIZE R(I), TRANSFORM, STORE TRANSFORMED VALUE IN Y(I), I = 1,4.
      DO 115 I = 1,3,2
      R(I) = SQRT(-2.*ALOG(R(I)))
      J = I + 1
      R(J) = 6.2831853*R(J)
      Y(I) = GVAV(I) + GVSD(I)*R(I)*COS(R(J))
  115 Y(J) = GVAV(J) + GVSD(J)*R(I)*SIN(R(J))
C
```

```
C ADJUST A,B,C,D FOR COMMON ELEMENTS, IF REQUIRED                      RTCR2410
      IF(KMLMT.EQ.0) GO TO 125                                         RTCR2420
      KC = 0                                                           RTCR2430
      DO 120 J = 1,3                                                   RTCR2440
      JK = J + 1                                                       RTCR2450
      DO 120 K = JK,4                                                  RTCR2460
      KC = KC + 1                                                      RTCR2470
      IF (CMSD(KC)) 120,120,117                                        RTCR2480
  117 CALL UNNO(Q,RA)                                                  RTCR2490
      CALL UNNO(Q,RB)                                                  RTCR2500
      ZA = SQRT (-2.*ALOG(RA))                                         RTCR2510
      ZB = COS(6.2831853*RB)                                           RTCR2520
      CMEL = CMSD(KC)*ZA*ZB                                            RTCR2530
      Y(J) = Y(J) + CMEL                                               RTCR2540
C SIGN OF INCREMENT TO Y(K) DEPENDS ON SIGN OF RHO(KC)                 RTCR2550
      IF (RHO(KC)) 118,118,119                                         RTCR2560
  118 Y(K) = Y(K) - CMEL                                               RTCR2570
      GO TO 120                                                        RTCR2580
  119 Y(K) = Y(K) + CMEL                                               RTCR2590
  120 CONTINUE                                                         RTCR2600
C                                                                      RTCR2610
C STORE NUMERATORS OF RATIOS IN Y(N),N=5,16                            RTCR2620
  125 I = 4                                                            RTCR2630
      DO 130 J = 1,3                                                   RTCR2640
      DO 130 K = 1,4                                                   RTCR2650
      I = I + 1                                                        RTCR2660
  130 Y(I) = Y(K)                                                      RTCR2670
C                                                                      RTCR2680
C DIVIDE NUMERATORS BY APPROPRIATE DENOMINATORS                        RTCR2690
      K = 17                                                           RTCR2700
      DO 140 I = 1,4                                                   RTCR2710
      DO 140 J = 1,3                                                   RTCR2720
      K = K - 1                                                        RTCR2730
  140 Y(K) = Y(K)/Y(I)                                                 RTCR2740
C                                                                      RTCR2750
```

```
C CUMULATE SUMS OF VARIABLES, SQUARES AND X-PRODUCTS FOR ONE ITEM         RTCR2760
      DO 145 K = 1,16                                                      RTCR2770
      SUM(K) = SUM(K) + Y(K)                                               RTCR2780
      DO 145 J = K,16                                                      RTCR2790
  145 SIG(J,K) = SIG(J,K) + Y(J)*Y(K)                                      RTCR2800
C                                                                          RTCR2810
C STORE AVERAGES IN AV, CONVERT SIG TO COV MATRIX                          RTCR2820
      RN = NR                                                              RTCR2830
      RNL = RN - 1.                                                        RTCR2840
      DO 150 K = 1,16                                                      RTCR2850
      AV(K) = SUM(K)/RN                                                    RTCR2860
      DO 150 J = K,16                                                      RTCR2870
  150 SIG(J,K) =(SIG(J,K) - SUM(J)*AV(K))/RNL                              RTCR2880
C                                                                          RTCR2890
C MAKE DIAG. ELEMENTS STD. DEVIATIONS, OFF-DIAG. CORR. COEFFICIENTS        RTCR2900
      DO 160 I = 1,16                                                      RTCR2910
  160 SIG(I,I) = SQRT(SIG(I,I))                                            RTCR2920
      DO 170 J = 1,15                                                      RTCR2930
      K = J + 1                                                            RTCR2940
      DO 170 I = K,16                                                      RTCR2950
      SIG(I,J) = SIG(I,J)/(SIG(I,I)*SIG(J,J))                              RTCR2960
  170 SIG(J,I) = SIG(I,J)                                                  RTCR2970
C                                                                          RTCR2980
C COMPUTATIONS COMPLETE.  PREPARE TO PRINT                                 RTCR2990
C                                                                          RTCR3000
      IF (KMLMT.EQ.0) GO TO 200                                            RTCR3010
C ADJUST PARAMTER VALUES FOR COMMON ELEMENT EFFECTS                        RTCR3020
      KC = 0                                                               RTCR3030
      DO 190 J = 1,3                                                       RTCR3040
      KJ = J + 1                                                           RTCR3050
      DO 190 K = KJ,4                                                      RTCR3060
      KC = KC + 1                                                          RTCR3070
      IF(CMSD(KC)) 190,190,180                                             RTCR3080
  180 GVSD(J) = SQRT(GVSD(J)**2 + CMSD(KC)**2)                             RTCR3090
      GVSD(K) = SQRT(GVSD(K)**2+CMSD(KC)**2)                               RTCR3100
```

```
  190 CONTINUE                                                          RTCR3110
C                                                                       RTCR3120
C PRINT COMPARISON OF PARENT AND SAMPLE RATIO TERMS                     RTCR3130
  200 WRITE (OUT,30)                                                    RTCR3140
      WRITE (OUT,8)(NAM(J),GVAV(J),AV(J),GVSD(J),SIG(J,J),J=1,4)        RTCR3150
      LINE = LINE + 15                                                  RTCR3160
C                                                                       RTCR3170
      IF (ROST(1).EQ.0) GO TO 220                                       RTCR3180
C PRINT COMPLETE MATRIX OF CORRELATIONS AND STD.-DEVS. IF REQUESTED     RTCR3190
      WRITE (OUT,10) NAM                                                RTCR3200
      DO 210 J = 1,16                                                   RTCR3210
  210 WRITE (OUT,12) NAM(J), (SIG(I,J),I=1,16)                          RTCR3220
      LINE = LINE + 19                                                  RTCR3230
C                                                                       RTCR3240
C PACK AND LIST CORRELATIONS BETWEEN TERMS OF RATIOS                    RTCR3250
  220 IF (ROST(2).EQ.0) GO TO 235                                       RTCR3260
      IP = 0                                                            RTCR3270
      N = 0                                                             RTCR3280
      DO 230 J = 1,3                                                    RTCR3290
      K = J + 1                                                         RTCR3300
      DO 230 I = K,4                                                    RTCR3310
      IP = IP + 1                                                       RTCR3320
      PRBF(IP) = NAM(J)                                                 RTCR3330
      IP = IP + 1                                                       RTCR3340
      PRBF(IP) = NAM(I)                                                 RTCR3350
      N = N + 1                                                         RTCR3360
  230 PRBF(12+N) = SIG(I,J)                                             RTCR3370
      WRITE (OUT,14) (PRBF(N),N=1,18)                                   RTCR3380
      LINE = LINE + 4                                                   RTCR3390
C                                                                       RTCR3400
  235 IF (ROST(3).EQ.0) GO TO 245                                       RTCR3410
C PACK AND LIST CORRELATIONS BETWEEN RATIOS AND THEIR NUMERATORS        RTCR3420
      KLO = 4                                                           RTCR3430
      KP = 0                                                            RTCR3440
      N = 0                                                             RTCR3450
```

```
      DO 240 J = 1,4                                              RTCR3460
      KLO = KLO + 1                                               RTCR3470
      DO 240 K = KLO,16,4                                         RTCR3480
      KP = KP + 1                                                 RTCR3490
      PRBF(KP) = NAM(K)                                           RTCR3500
      KP = KP + 1                                                 RTCR3510
      PRBF(KP) = NAM(J)                                           RTCR3520
      N = N + 1                                                   RTCR3530
  240 PRBF(24+N) = SIG(K,J)                                       RTCR3540
      LINE = LINF + 5                                             RTCR3550
      IF (55 - LINF) 243,244,244                                  RTCR3560
  243 WRITE (OUT,NUPG)                                            RTCR3570
      LINE = 0                                                    RTCR3580
  244 WRITE (OUT,16) (PRBF(N), N = 1,36)                          RTCR3590
      LINE = LINF + 5                                             RTCR3600
  245 IF(RQST(4).EQ.0) GO TO 255                                  RTCR3610
C                                                                 RTCR3620
C PACK AND LIST CORRELATIONS BETWEEN RATIOS AND THEIR DENOMINATORS RTCR3630
      KLO = 17                                                    RTCR3640
      KP = 0                                                      RTCR3650
      N = 0                                                       RTCR3660
      DO 250 J = 1,4                                              RTCR3670
      KHI = KLO- 1                                                RTCR3680
      KLO = KLO - 3                                               RTCR3690
      DO 250 K = KLO,KHI                                          RTCR3700
      KP = KP + 1                                                 RTCR3710
      PRBF(KP) = NAM(K)                                           RTCR3720
      KP = KP + 1                                                 RTCR3730
      PRBF(KP) = NAM(J)                                           RTCR3740
      N = N + 1                                                   RTCR3750
  250 PRBF(24+N) = SIG(K,J)                                       RTCR3760
      LINE = LINE + 5                                             RTCR3770
      IF (55 - LINE) 253,254,254                                  RTCR3780
  253 WRITE (OUT,NUPG)                                            RTCR3790
      LINE = 0                                                    RTCR3800
```

```
  254 WRITE(OUT,18) (PRBF(N),N=1,36)
C
  255 IF(RQST(5).EQ.0) GO TO 280
C PACK AND LIST CORRELATIONS BETWEEN RATIOS WITH COMMON NUMERATORS
      INK = 1
      KP = 0
      N = 0
      I = 5
      J = 13
      K = 9
  260 KP = KP + 1
      PRBF(KP) = NAM(I)
      KP = KP + 1
      PRBF(KP) = NAM(K)
      N = N + 1
      PRBF(24+N) = SIG(I,K)
      KP = KP + 1
      PRBF(KP) = NAM(I)
      KP = KP + 1
      PRBF(KP) = NAM(J)
      N = N + 1
      PRBF(24+N) = SIG(I,J)
      KP = KP + 1
      PRBF(KP) = NAM(K)
      KP = KP + 1
      PRBF(KP) = NAM(J)
      N = N + 1
      PRBF(24+N) = SIG(K,J)
      I = I + INK
      J = J + INK
      K = K + INK
      IF (KP.LT.24) GO TO 260
      IF(INK - 2) 275,275,270
C WRITE CORR. WITH COMMON DENOMIN., RETURN FOR NEW COMMAND
  270 LINE = LINE + 5
```

```
      IF (55 - LINE) 2700,2701,2701
 2700 WRITE (OUT,NUPG)
      LINE = 0
 2701 WRITE (OUT,22) (PRBF(N),N=1,36)
C
  271 IF(RQST(7).EQ.0) GO TO 272
C PRINT CORRELATIONS IN WHICH THE SAME TERM IS THE NUMERATOR OF ONE
C AND THE DENOMINATOR OF THE OTHER.
      LINE = LINE + 5
      IF (55 - LINE) 2710,2711,2711
 2710 WRITE (OUT,NUPG)
      LINE = 0
 2711 WRITE (OUT,24) NAM(1), NAM(2), NAM(14), NAM(5), NAM(14), NAM(9),
     *NAM(15), NAM(5), NAM(15), NAM(13), NAM(16), NAM(9), NAM(16),
     *NAM(13), NAM(11), NAM(6), NAM(11), NAM(14), NAM(12), NAM(10),
     *NAM(12), NAM(14), NAM(13), NAM(6), NAM(13), NAM(10), SIG(14,5),
     *SIG(14,9), SIG(15,5), SIG(15,13), SIG(16,9), SIG(16,13), SIG(11,6)
     *, SIG(11,14), SIG(12,10), SIG(12,14), SIG(13,6), SIG(13,10)
C
      LINE = LINE + 5
      IF (55 - LINE) 2712,2713,2713
 2712 WRITE (OUT,NUPG)
      LINE = 0
 2713 WRITE (OUT,24) NAM(3), NAM(4), NAM(8), NAM(11), NAM(8), NAM(15),
     *NAM(9), NAM(7), NAM(9), NAM(11), NAM(10), NAM(7), NAM(10), NAM(15)
     *, NAM(5), NAM(8), NAM(5), NAM(12), NAM(6), NAM(8), NAM(6), NAM(16)
     *, NAM(7), NAM(12), NAM(7), NAM(16), SIG(8,11), SIG(8,15),
     *SIG(9,7), SIG(9,11), SIG(10,7), SIG(10,15), SIG(5,8), SIG(5,12),
     *SIG(6,8), SIG(6,16), SIG(7,12), SIG(7,16)
C
  272 IF(RQST(8).EQ.0) GO TO 273
C PRINT CORRELATIONS OF RATIOS WHICH ARE RECIPROCALS OF EACH OTHER.
      LINE = LINE + 4
      IF (55 - LINE) 2720,2721,2721
 2720 WRITE (OUT,NUPG)
      LINE = 0
```

```
 2721 WRITE (OUT,26) NAM(13),NAM(14),NAM(9),NAM(15),NAM(5),NAM(16),NAM(1
     *0),NAM(11),NAM(6),NAM(12),NAM(7),NAM(8),SIG(13,14),SIG(9,15),SIG(5
     *,16),SIG(10,11),SIG(6,12),SIG(7,8)
  273 IF (RQST(9).EQ.0) GO TO 101
C PRINT CORRELATIONS BETWEEN RATIOS LACKING COMMON TERMS
      LINE = LINE + 5
      IF (55 - LINE) 2730,2731,2731
 2730 WRITE (OUT,NUPG)
      LINE = 0
 2731 WRITE (OUT,34) NAM(5), NAM(10), NAM(16), NAM(10), NAM(5), NAM(11),
     *NAM(16), NAM(11), NAM(6), NAM(9), NAM(12), NAM(9), NAM(6), NAM(15)
     *, NAM(12), NAM(15), NAM(7), NAM(13), NAM(8), NAM(13), NAM(7),
     * NAM(14), NAM(8), NAM(14), SIG(5,10), SIG(16,10), SIG(5,11),
     *SIG(16,11), SIG(6,9), SIG(12,9), SIG(6,15), SIG(12,15), SIG(7,13),
     *SIG(8,13), SIG(7,14), SIG(8,14)
      GO TO 101
C WRITE CORR. WITH COMMON NUMERATORS, CONTINUE
  275 LINE = LINE + 5
      IF (55 - LINE) 2750,2751,2751
 2750 WRITE (OUT,NUPG)
      LINE = 0
 2751 WRITE (OUT,20) (PRBF(N),N=1,36)
  280 IF (RQST(6).EQ.0) GO TO 271
C INITIALIZE POINTERS TO PACK CORR. BETWEEN RATIOS WITH COMM. DENOMIN.
      INK = 3
      KP = 0
      N = 0
      I = 5
      J = 7
      K = 6
      GO TO 260
C
C RECORD CURRENT VALUE OF RANDOM NUMBER GENERATOR SEED.
  300 WRITE(OUT,28)Q
      STOP
      END
```

APPENDIX 2

```
C                           PROGRAM RANEX
C EXAMINES RANDOMNESS AND UNIFORMITY OF A SEQUENCE OF NUMBERS BY '
C    1. COUNTING FREQUENCIES OF RUNS OF LENGTH K OVER OR UNDER 1/2,
C    2. COUNTING FREQUENCIES OF RUNS OF LENGTH M UP OR DOWN.
C    3. COUNTING FREQUENCIES WITH WHICH PAIRED NUMBERS SEPARATED BY A
C GIVEN DISTANCE (LAG) FALL IN JOINT SIZE CLASS(M,N), M : N = 1,20. THIS
C TABULATION IS OMITTED WHEN SAMPLES OF LESS THAN 2000 NUMBERS ARE USED.
C    4. COMPUTING CHI-SQUARE(S) ON ASSUMPTION THAT NUMBERS  AND NUMBER-
C PAIRS ARE UNIFORMLY DISTRIBUTED,
C    5. COMPUTING AUTO-CORRELATION COEFFICIENTS FOR LAGS J, J=1,LAG
C
C INPUT IS FROM SUBROUTINE UNNO(L,F) WHERE L IS A RANDOM INTEGER
C AND F IS A FLOATED RANDOM NUMBER IN THE RANGE (0,1)
C
C PROGRAM WRITTEN BY F. CHAYES FOR NSF STATISTICAL GEOLOGY INSTITUTE,
C                       CHICAGO CIRCLE, 1972.
C
C ************************************************************************
C *                                                                      *
C *                         CARD INPUT TO RANEX                          *
C *                                                                      *
C *     COL.       VARIABLE             FUNCTION OR DEFINITION           *
C *                                                                      *
C *   TITLE CARD                                                (20A4)   *
C *      1-80         TITL      TITLE INFORMATION, 80 CHARACTERS         *
C *                                                                      *
C *   COMMAND CARD                                           (3I5,I15)   *
C *      1-5          NOKMP     NUMBER OF ITEMS PER SAMPLE               *
C *      6-10         ITER      NUMBER OF SAMPLES                        *
C *     11-15         LAG       INTERVAL BETWEEN MEMBERS OF A PAIR (=0)  *
```

```
C *     16-30          KRNL      INITIAL VALUE OF RANDOM NUMBER SEED       *
C *                              ( K; USE K                                *
C *                                 0; USE 1ST RESIDUE FROM GENERATOR      *
C *                                -1; USE LAST VALUE REACHED ON PRE-      *
C *                                    CEEDING PASS OF SAME EXECUTION)     *
C *                                                                        *
C ************************************************************************
C
      DIMENSION CHIPR(20), KL(20,20), LN1(4), LN2(4), LN3(4), LN4(4),
     *LN5(4), LN6(4), NRDN(20), NRGT(20), NRLS(20), NRUP(20), NUMF(20),
     *RLOU(20), RLUD(20), RNVEC(100), TITL(20), UKLM(20), AUCR(200)
C
      DOUBLE PRECISION AUCR,OKMP
      INTEGER OUT
C
      EQUIVALENCE (RNVEC,CHIPR), (RNVEC(21),NUMF), (RNVEC(41),RLOU),
     *(RNVEC(61),RLUD)
C
      DATA IN,OUT/5,6/
      DATA UKLM/.05,.10,.15,.20,.25,.30,.35,.40,.45,.50,.55,.60,.65,.70,
     *.75,.80,.85,.90,.95,1./
      DATA LN1,LN2,LN3,LN4/4H RUN,4HS OF,4H LEN,3HGTH,4H  NU,4HMBER,
     *4HS + ,3H1/2,4H  NU,4HMBER,4HS = ,3H1/2,4H    ,4H TOT,4HAL  ,
     *3H  /
      DATA LN5,LN6/4HORUN,4HS DO,4HWN  ,3H   ,4H RUN,4HS UP,4H    ,3H
     */
C
C                              INPUT FORMATS
    1 FORMAT (20A4/3I5,I15)
C
C                              OUTPUT FORMATS
    2 FORMAT ('1',20A4,2X,'LAG =',I4,', LENGTH =',I6,', START =',I13/115
     *X,'SAMPLE NO.',I3//)
    4 FORMAT (3A4,1A3,13I7,4I5/17X,3I5)
    6 FORMAT (///30X,'JUNCOSA ARRAY' - FREQUENCIES OF PAIRS IN SIZE CLAS
```

```
  *SES R AND C'//56X,'RANDOM NO.(K +',I3,') +'/15X,20F5.2/' RANDOM NO
  *. K +')
 8 FORMAT (10X,F5.2,20I5)
10 FORMAT (/6X,'CHI-SQ.',2X,20F5.1//)
12 FORMAT  ('OCHI-SQUARE FOR DEPARTURE FROM UNIFORM DISTRIBUTION OF N
  *UMBERS IN RANGE (0,1) IS ',F10.3,'.')
14 FORMAT (/45X,'AUTOCORRELATION OVER INDICATED LAG DISTANCE(S) -')
16 FORMAT (4X,'F(LAG NOS.)',20I5)
18 FORMAT (5X,'EXPECTED',F11.1,12F7.1,4F5.1/19X,3F5.1)
20 FORMAT ('1*****REQUEST FOR CALCULATIONS WITH LAG = 0 IGNORED*****'
  *)
22 FORMAT (10(I3,F7.4,',  '))
24 FORMAT (/30X,'FINAL RANDOM NUMBER IS ',I15/1H1)
26 FORMAT ('1 ERROR' PROGRAM SKIPS PROBLEM INITIATED BY FAULTY INPUT
  *CARD, SEARCHES FOR TITLE CARD OF NEXT PROBLEM.')
28 FORMAT (1H0,20X,'SAMPLE TOO SMALL (+ 2000) FOR EFFECTIVE LAG PAIR
  *COMPARISONS. PROCEED TO UNIFORMITY TEST.'//56X,'RANDOM NO.(K +',I3
  *,')+'/15X,20F5.2)
C
C READ COMMAND CARD
60 READ (IN,1,ERR=63,END=350) TITL,NOKMP,ITER,LAG,KRNL
   IF (ITER.EQ.0) ITER = 1
   IF (LAG.GT.0) GO TO 65
   WRITE (OUT,20)
63 WRITE (OUT,26)
   GO TO 60
C
C INITIALIZE RANDOM NUMBER GENERATOR
65 IF(KRNL) 70,80,90
70 KR = KRLST
   KRNL = KRLST
   GO TO 100
80 CALL UNNO(KR,RA)
   KRNL = KR
   GO TO 100
```

```
   90 KR = KRNL
C
      ITR = 0
  100 ITR = ITR + 1
      IF (ITR.GT.ITER) GO TO 60
      IF (ITR.GT.1) KRNL = KRLST
C GENERATE AND STORE (LAG+1) RANDOM NUMBERS
      L = ABS(LAG) + 1
      DO 110 J = 1,L
  110 CALL UNNO(KR,RNVEC(J))
      IF (L.LT.2) CALL UNNO(KR,RNVEC(2))
C
C CLEAR ALL FREQUENCY AND OTHER CUMULATORS
      DO 115 I = 1,20
      NRGT(I) = 0
      NRLS(I) = 0
      NRUP(I) = 0
      NRDN(I) = 0
      DO 115 J = 1,20
  115 KL(J,I) = 0
      DO 120 J = 1,200
  120 AUCR(J) = 0.0
C
C RECORD RUNS UP-DOWN, OVER-AND UNDER 1/2, IN FIRST (LAG+1)NUMBERS
      M = 1
      K = 1
      MM = 0
      KK = 0
      DF = RNVEC(2) - RNVEC(1)
      KF = RNVEC(1) + 0.5
      DO 175 J = 2,L
      IF (J.LT.3) GO TO 158
      DL = RNVEC(J) - RNVEC(J-1)
      IF (DF) 125,125,140
C                 CURRENT RUN IS DOWN
```

```
  125 IF(DL) 130,130,135                                                  RNEX1360
  130 M = M + 1                                                           RNEX1370
      IF (M.EQ.20) M = 20                                                 RNEX1380
      IF (M.GT.MM) MM = M                                                 RNEX1390
      GO TO 155                                                           RNEX1400
  135 NRDN(M) = NRDN(M) + 1                                               RNEX1410
      GO TO 150                                                           RNEX1420
C                    CURRENT RUN IS UP                                    RNEX1430
  140 IF(DL) 145,145,130                                                  RNEX1440
  145 NRUP(M) = NRUP(M) + 1                                               RNEX1450
  150 M = 1                                                               RNEX1460
  155 DF = DL                                                             RNEX1470
  158 KS = RNVEC(J) + 0.5                                                 RNEX1480
      IF (KS + KF - 1) 160,165,160                                        RNEX1490
C                    CURRENT PAIR THE SAME, INCREMENT COUNTER             RNEX1500
  160 K = K + 1                                                           RNEX1510
      IF (K.GT.20) K = 20                                                 RNEX1520
      IF (K.GT.KK) KK = K                                                 RNEX1530
      GO TO 175                                                           RNEX1540
C                    CURRENT PAIR DIFFER, RECORD RUN LENGTH, RESET K      RNEX1550
  165 IF(KF.EQ.1) GO TO 170                                               RNEX1560
      NRLS(K) = NRLS(K) + 1                                               RNEX1570
      GO TO 173                                                           RNEX1580
  170 NRGT(K) = NRGT(K) + 1                                               RNEX1590
  173 K = 1                                                               RNEX1600
  175 KF = KS                                                             RNEX1610
C                                                                         RNEX1620
C CLASSIFY AND STORE FREQUENCIES OF PAIRED NUMBERS M AND M+L, CUMULATE    RNEX1630
C PRODUCTS FOR AUTO-CORRELATION COEFFICIENT, MAINTAIN COUNT OF RUNS OF    RNEX1640
C LENGTH K OVER AND UNDER 1/2.                                            RNEX1650
C                                                                         RNEX1660
      DO 300 N = 1,NOKMP                                                  RNEX1670
C CUMULATE AUTO-CORR. SUM OF PRODUCTS                                     RNEX1680
      LL = L - 1                                                          RNEX1690
      DO 178 I = 1,LL                                                     RNEX1700
```

```
  178 AUCR(I) = AUCR(I) + RNVEC(L-I)*RNVEC(L)
C CLASSIFY CURRENT NUMBER-PAIR RNVEC(1) AND RNVEC(L)
      DO 190 I = 1,20
      IF (RNVEC(1) - UKLM(I)) 180,180,190
  180 IC = I
      GO TO 200
  190 CONTINUE
      IC = 20
  200 DO 220 I = 1,20
      IF  (RNVEC(L) - UKLM(I)) 210,210,220
  210 IR = I
      GO TO 230
  220 CONTINUE
      IR = 20
  230 KL(IR,IC) = KL(IR,IC) + 1
C
C SHIFT ELEMENTS OF RNVEC DOWN ONE CELL, GENERATE NEW RNVEC(L)
      DO 240 JR = 2,L
  240 RNVEC(JR - 1) = RNVEC(JR)
      CALL UNND(KR,RNVEC(L))
C
C CONTINUE COUNT OF RUNS UP-DOWN, OVER-UNDER 1/2
      DL = RNVEC(L) - RNVEC(L-1)
      IF (DF) 245,245,260
  245 IF(DL) 250,250,255
  250 M = M + 1
      IF (M.EQ.20) M = 20
      IF (M.GT.MM) MM = M
      GO TO 275
  255 NRDN(M) = NRDN(M) + 1
      GO TO 270
  260 IF(DL) 265,265,250
  265 NRUP(M) = NRUP(M) + 1
  270 M = 1
  275 DF = DL
```

```
      KS = RNVEC(L) + 0.5                                               RNEX2060
      IF (KS + KF - 1) 280,285,280                                      RNEX2070
  280 K = K + 1                                                         RNEX2080
      IF (K.GT.20) K = 20                                               RNEX2090
      IF (K.GT.KK) KK = K                                               RNEX2100
      GO TO 300                                                         RNEX2110
  285 IF (KF.EQ.1) GO TO 290                                            RNEX2120
      NRLS(K) = NRLS(K) + 1                                             RNEX2130
      GO TO 295                                                         RNEX2140
  290 NRGT(K) = NRGT(K) + 1                                             RNEX2150
  295 K = 1                                                             RNEX2160
  300 KF = KS                                                           RNEX2170
C                     RECORD LAST RUN OF EACH TYPE                      RNEX2180
      IF(DL) 302,302,304                                                RNEX2190
  302 NRDN(M) = NRDN(M) + 1                                             RNEX2200
      GO TO 306                                                         RNEX2210
  304 NRUP(M) = NRUP(M) + 1                                             RNEX2220
  306 IF(KS.GT.0) GO TO 308                                             RNEX2230
      NRLS(K) = NRLS(K) + 1                                             RNEX2240
      GO TO 309                                                         RNEX2250
  308 NRGT(K) = NRGT(K) + 1                                             RNEX2260
C                                                                       RNEX2270
C CALCULATE AUTO-CORRELATION COEFF., BEGIN WRITING                      RNEX2280
  309 KRLST = KR                                                        RNEX2290
      OKMP = NOKMP                                                      RNEX2300
      DO 3095 J = 1,LL                                                  RNEX2310
 3095 AUCR(J) = AUCR(J)/OKMP                                            RNEX2320
      WRITE (OUT,2) TITL,LAG,NOKMP,KRNL,ITR                             RNEX2330
C                                                                       RNEX2340
C CALC. OBS. : EXP. NUMBERS OF RUNS OF LENGTH + : = 1/2                 RNEX2350
      DO 310 J = 1,KK                                                   RNEX2360
  310 RLOU(J) = FLOAT(NOKMP + LAG - J - 4)/2.**(J+1)                    RNEX2370
C COMPUTE CHI-SQUARES FOR INDIVIDUALS AND PAIRED FREQUENCIES            RNEX2380
      CHIFR = 0.0                                                       RNEX2390
      TMP = FLOAT(NOKMP)/400.                                           RNEX2400
```

```
      TMF = 20.*TMP                                                     RNEX2410
      DO 330 IC = 1,20                                                  RNEX2420
      CHIPR(IC) = 0.                                                    RNEX2430
      SUMFR = 0.                                                        RNEX2440
      NUMF(IC) = 0                                                      RNEX2450
      DO 320 IR = 1,20                                                  RNEX2460
      CLFR = KL(IR,IC)                                                  RNEX2470
      SUMFR = SUMFR + CLFR                                              RNEX2480
      NUMF(IC) = NUMF(IC) + KL(IR,IC)                                   RNEX2490
  320 CHIPR(IC) = CHIPR(IC) + (CLFR - TMP)**2                           RNEX2500
      CHIPR(IC) = CHIPR(IC)/TMP                                         RNEX2510
  330 CHIFR = CHIFR + (SUMFR- TMF) ** 2                                 RNEX2520
      CHIFR = CHIFR/TMF                                                 RNEX2530
C                                                                       RNEX2540
C COMPLETE WRITING                                                      RNEX2550
      WRITE (OUT,4) LN1, (J,J=1,KK)                                     RNEX2560
      WRITE (OUT,4) LN2, (NRLS(J),J=1,KK)                               RNEX2570
      WRITE (OUT,4) LN3, (NRGT(J),J=1,KK)                               RNEX2580
      DO 332 J = 1,KK                                                   RNEX2590
  332 NRGT(J) = NRGT(J) + NRLS(J)                                       RNEX2600
      WRITE (OUT,4) LN4, (NRGT(J),J=1,KK)                               RNEX2610
      WRITE (OUT,18) (RLOU(J), J = 1,KK)                                RNEX2620
      WRITE (OUT,4) LN5,(NRDN(J),J=1,MM)                                RNEX2630
      WRITE (OUT,4) LN6,(NRUP(J),J = 1,MM)                              RNEX2640
C                                                                       RNEX2650
C CALC. OBS. AND EXP. NUMBERS OF RUNS OF LENGTH K 'UP-AND-DOWN'.        RNEX2660
      BOT = 6.                                                          RNEX2670
      DO 335 J = 1,MM                                                   RNEX2680
      FM = J + 3                                                        RNEX2690
      BOT = BOT * FM                                                    RNEX2700
      RLUD(J) = 2*((J**2+3*J+1)*(NOKMP+LAG+1)-(J**3+3*J**2-J-4))        RNEX2710
      RLUD(J) = RLUD(J)/BOT                                             RNEX2720
  335 NRUP(J) = NRDN(J) + NRUP(J)                                       RNEX2730
      WRITE (OUT,4) LN4,(NRUP(J),J=1,MM)                                RNEX2740
      WRITE (OUT,18) (RLUD(J),J=1,MM)                                   RNEX2750
```

```
    IF (NOKMP.LT.2000) GO TO 345
    WRITE (OUT,6) LAG, (UKLM(I),I=1,20)
    DO 340 J = 1,20
340 WRITE (OUT,8) UKLM(J),(KL(J,K),K=1,20 )
    WRITE (OUT,10) (CHIPR(J),J=1,20)
345 IF (NOKMP.LT.2000) WRITE (OUT,28) LAG,UKLM
    WRITE (OUT,16) NUMF
    WRITE (OUT,12) CHIFR
    WRITE (OUT,14)
    WRITE (OUT,22) (J,AUCR(J),J=1,LL)
    WRITE (OUT,24) KRLST
    GO TO 100
350 STOP
    END
    SUBROUTINE UNNO(L,F)
       PSEUDO-RANDOM NUMBER GENERATOR      CODED BY L.W.FINGER
          CALLING SEQUENCE IS CALL UNNO (L,F)
              L IS THE LAST RANDOM NUMBER CALCULATED. IF NOT CHANGED,
                 THE NUMBERS COME OUT IN SEQUENCE.
             F IS THE RANDOM NUMBER IN REAL FORM AND IS UNIFORMLY
                DISTRIBUTED IN THE RANGE (0,1)
    DOUBLE PRECISION D
    IF(L.EQ.0)L=23192344
    D=L
    D=DMOD(513.0D0*D,2147483647.0D0)
    L=D
    L=L+1
    F=(D+1.0D0)/2147483647.0D0
    RETURN
    END
```

Chapter 6

Computer Perspectives in Geology

Daniel F. Merriam

6.1 GENERALITIES

The introduction of computers into society in the mid-19th century ushered in the space-age and, with it, some special problems. Although the idea of computers and computing has been with us for some time, the explosive development of computers just after World War II was not foreseen, and the public as a whole was not ready for such dramatic and rapid changes in this field. So, although we have seen men on the moon, world-wide weather forcasting via satellites, and breakthroughs in medicine and science, some of the 1984 predictions have come to pass. There has been an invasion of privacy, many trades and practices have been declared obsolete and redundant, and an impersonal touch has been added to an ever-increasing complex society.

The problems that plague the general public also plague science, and geology is no exception. The numerous changes stemming from the computer have forced geologists to reevaluate their contributions and in some instances, the results have been startling. Geologists have been put in the position of having to think and formulate their problems and methods of solution in much greater detail. Instead of replacing geologists, the computer has created a demand for more and better trained ones. One effect the computer has had on geology is to force a metamorphosis on us, that is a change from a qualitative science to a quantitative one. Purists claim even that geology is only now becoming a science.

6.2 EARLY BEGINNINGS

Computers date from 1812 when Charles Babbage invented his difference machine. The machine was designed to operate automatically without human intervention. Although the underlying principle was simple, Babbage encountered almost insurmountable problems in building his machine. It was unfortunate that Babbage was unable to complete his marvelous invention, but he was too far ahead of his time, and technology simply had not developed to the necessary extent.

The next important event was the development of a punched-card system by Herman Hollerith in about 1890. Mr. Hollerith worked for the U.S. Bureau of Census, and he recognized early the need for the rapid manipulation of data. Even then it was difficult to summarize census information. Hollerith's original idea of the punched card is probably the device most used today for input to the computer.

During World War II the development of computers was accelerated, especially to aid in solving problems in ballistic missile development and in the development of the atomic bomb. Rapid strides were made possible through the adaptation of Boolean algebra which had been formulated some years before by Claude Shannon. Although the first computers were awkward to use and slow by today's standards, they served their purpose and laid the foundation for a complex and dynamic industry.

6.3 USAGE IN GENERAL

Computers may be used for a number of reasons, including (1) saving time and effort, (2) making use of information in ways that would be virtually impossible without the aid of computers, and (3) improving the rigor of thought processes (Harbaugh and Merriam, 1968). The application of computer techniques to solving problems in the earth sciences is now important and becoming more so. Many of these applications were unthought of just a few years ago and indeed just a few months or even weeks ago. As aptly stated by P. C. Hammer (1966, personal communication)

"People who are thinking about what they are doing are using computers."

Although some of the techniques were possible before the advent of the computer, many were not. Execution is presently feasible only because of the ease and speed with which they can be accomplished and as stated by P. C. Hammer (1966, personal communication)

"A computer is an intelligence amplifier."

The ability to save time and effort and to some extent the ability to manipulate data in ways that are impossible otherwise are mainly an aspect of computers and computing systems themselves. It is essential, therefore, that some personal involvement in the process of computing be attained.

Two of the most important aspects of the use of computers are repeatability and reliability; that is, anyone can take a set of data and reproduce the results within the same limits of accuracy by the same method. It is not possible to do this with qualitative methods and these two aspects can not be overestimated (Merriam, 1965).

Obviously there are times and places when and where a computer can be used to good advantage. These are (1) if there is a large volume of data, (2) if speed or frequency is necessary in retrieval of data, or (3) if a particular problem is extremely complex. There is an area between these extremes where it is easier to do the required manipulations manually.

There are other considerations on whether to use a computer. These include (1) the availability of programs, (2) the reducibility of the data to numeric form, and (3) ease of accessibility and economic feasibility.

If it is necessary to develop programs, it can be extremely tedious and expensive. Obviously, if there is a low volume of data, or speed is no object, or the problem is not too complicated, it would be desirable to obtain results manually. Fortunately, however, programs are available from many sources and many may be adapted for a particular use.

Many geologic data are qualitative and not amenable to computer analysis. Many data also are incomplete or of poor quality and essentially useless by today's standards. It may be desirable in many instances simply to recollect the data. Another requirement is the necessity that the problem be expressed in a sequence of relatively simple, logical, algebraic statements. This is necessary because operations to manipulate the data must be explicit, precise, and unambiguous. This requirement can usually be met although it may take considerable thinking and planning on the part of the investigator.

6.4 USAGE IN GEOLOGY

Just as computers date from 1812, modern geology dates from the late 18th century and the work of James Hutton, a Scot. It is interesting to note that these events both took place in Britain at about the same time and about 150 years before the application of computers to geologic problems.

The first earth scientists to use computers were those who were numerically inclined. It is not surprising then that geophysicists were the first. They had been using slide rules and desk calculators for many years, and it was natural to adapt to a new and better method of processing their data. Other exceptions were those conducting statistical studies of sediments and their contained fossils and those working with engineering aspects, such as hydrologists. Where large quantities of data were handled, techniques were needed in manipulating them. Just as Herman Hollerith needed assistance with his census data, geologists needed help with their data processing. It is logical then that sorting of oil and gas data and stratigraphic information was accomplished with punch cards in the early 1950's (Parker, 1952). Early bibliographic systems also utilized punched cards. As data accumulated, it was necessary to sort faster, and when computers became generally available, automatic procedures replaced manual ones.

It might be imagined that because computers have only been commercially available for about 18 years that the utilization of them by geologists has been most recent. This is indeed true, and only in the past 10 years has this involvement become increasingly important. The importance can be judged by the number of geologic publications appearing which have in some way utilized the computer as shown in Figure 6.1. The number of publications is increasing rapidly and an obvious increase occurred in the number of reports on research beginning in 1962, which is the result of the general availability of second-generation computers. Geology entered the computer age with a publication in a regularly issued geology journal of a geologically oriented IBM 650 program by Krumbein and Sloss (1958). The original program is reproduced in Figure 6.2. Other important events which have affected the development of computer applications in geology are listed in Table 6.1.

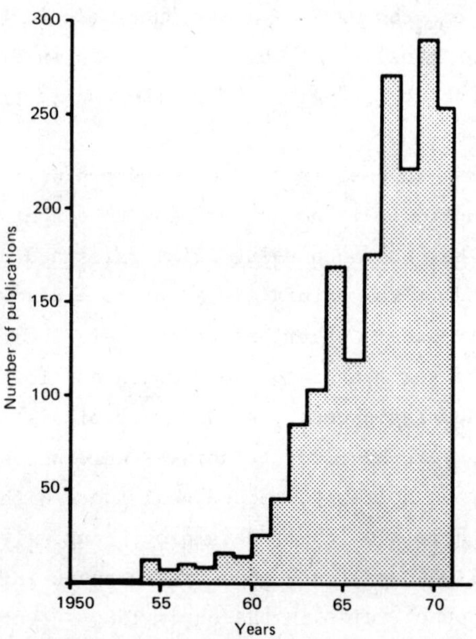

FIGURE 6.1

IBM 650 Basic Program for Three Percentages and Two Ratios

Zero drum, Start program at 0501, No Subroutines required

Location of Instruction	OP	Data	Instr.	Abbrev.	Remarks
0501	70	1501	0502	RD	Read data card
0502	65	1501	0503	RAL	Code in accumulator
0503	20	0727	0504	STL	Store code
0504	65	1503	0505	RAL	Total in accumulator
0505	20	0728	0506	STL	Store total thickness
0506	65	1504	0507	RAL	B to accumulator
0507	16	0660	0508	SL	Subtract 10 1010 1010
0508	45	0510	0509	BRNZ	Branch on B data
0509	24	0729	0515	STD	Store no data code
0510	15	0660	0511	AL	Add 10 1010 1010
0511	35	0004	0512	SLT	Shift left
0512	64	1503	0513	DVRU	Divide by total
0513	31	0001	0514	SRD	Shift and round
0514	20	0729	0515	STL	Store percent B
0515	65	1505	0516	RAL	C to accumulator
0516	16	0660	0517	SL	Subtract 10 1010 1010
0517	45	0519	0518	BRNZ	Branch on C data
0518	24	0730	0524	STD	Store no data code
0519	15	0660	0520	AL	Add 10 1010 1010
0520	35	0004	0521	SLT	Shift left
0521	64	1503	0522	DVRU	Divide by total
0522	31	0001	0523	SRD	Shift and round
0523	20	0730	0524	STL	Store percent C
0524	65	1506	0525	RAL	A to accumulator
0525	16	0660	0526	SL	Subtract 10 1010 1010
0526	45	0528	0527	BRNZ	Branch on A data
0527	24	0731	0552	STD	Store no data code
0528	15	0660	0529	AL	Add 10 1010 1010
0529	35	0004	0530	ALT	Shift left
0530	64	1503	0531	DVRU	Divide by total
0531	31	0001	0532	SRD	Shift and round
0532	20	0731	0552	STL	Store percent A
0552	71	0727	0553	PCH	Punch percentages
0553	65	1501	0554	RAL	Code in accumulator
0554	20	0827	0555	STL	Store code
0555	65	1503	0556	RAL	Total in accumulator
0556	20	0828	0557	STL	Store total
0557	65	1506	0558	RAL	A to accumulator
0558	45	0570	0559	BRNZ	Branch on nonzero A
0559	65	1504	0560	RAL	B to accumulator
0560	15	1505	0561	AL	Add C
0561	45	0564	0562	BRNZ	Branch on nonzero B + C
0562	65	0662	0563	RAL	Indeterminate code
0563	20	0829	0586	STL	Store code for 0/0
0564	16	0661	0563	SL	Subtract 20 2020 2020

FIGURE 6.2

FIGURE 6.2 (continued)

Location of Instruction	OP	Data	Instr.	Abbrev.	Remarks
0565	45	0567	0566	BRNZ	Branch on no B + C data
0566	24	0829	0586	STD	Store no data code
0567	65	0663	0568	RAL	Infinity code
0568	20	0829	0586	STL	Store infinity code
0569	16	0660	0570	SL	Subtract 10 1010 1010
0570	45	0572	0571	BRNZ	Branch on nonzero A
0571	24	0829	0586	STD	Store no data code
0572	65	1504	0573	RAL	B to accumulator
0573	16	0660	0574	SL	Subtract 10 1010 1010
0574	45	0576	0575	BRNZ	Branch on B data
0575	24	0829	0586	STD	Store no data code
0576	65	1505	0577	RAL	C to accumulator
0577	16	0660	0578	SL	Subtract 10 1010 1010
0578	45	0580	0579	BRNZ	Branch on C data
0579	24	0829	0586	STD	Store no data code
0580	65	1504	0581	RAL	B to accumulator
0581	15	1505	0582	AL	Add C
0582	35	0003	0583	SLT	Shift left
0583	64	1506	0584	DVRU	Divide by A
0584	31	0001	0585	SRD	Shift and round
0585	20	0829	0586	STL	Store ratio (B + C)/A
0586	65	1505	0587	RLA	C to accumulator
0587	45	0597	0588	BRNZ	Branch to C data
0588	65	1504	0589	RAL	B to accumulator
0589	45	0592	0590	BRNZ	Branch on B data
0590	65	0662	0591	RAL	Indeterminate code
0591	20	0830	0609	STL	Store indeterminate code
0592	16	0660	0593	SL	Subtract 10 1010 1010
0593	45	0595	0594	BRNZ	Branch on B data
0594	24	0830	0609	STD	Store no data code
0595	65	0663	0596	RAL	Infinity code
0596	20	0830	0609	STL	Store infinity code
0597	16	0660	0598	SL	Subtract 10 1010 1010
0598	45	0600	0599	BRNZ	Branch on C data
0599	24	0830	0609	STD	Store no data code
0600	65	1504	0601	RAL	B to accumulator
0601	16	0660	0602	SL	Subtract 10 1010 1010
0602	45	0604	0603	BRNZ	Branch on B data
0603	24	0830	0609	STD	Store no data code
0604	15	0660	0605	AL	Add 10 1010 1010
0605	35	0003	0606	SLT	Shift left
0606	64	1505	0607	DVRU	Divide by C
0607	31	0001	0608	SRD	Shift and round
0608	20	0830	0609	STL	Store ratio B/C
0609	71	0827	0501	PCH	Punch ratio card
0660	10	1010	1010		Const
0661	20	2020	2020		Const
0662	90	9090	9090		Const
0663	99	9999	9999		Const

TABLE 6.1 Important Events in Computer Applications in Geology

Year	Event
1812	Charles Babbage and his difference machine.
1890	Punched-card system developed by Herman Hollerith.
1941	Z3, first electronic computer, made in Germany.
1944	Mark I, the decimal electromechanical calculator put into operation at Harvard.
1946	ENIAC built at the University of Pennsylvania.
1951	UNIVAC, the first commercial computer.
1952	Digital plotters introduced.
1953	First FORTRAN compiler written.
1954	Introduction of the IBM 650, the first mass-produced computer.
1958	W. C. Krumbein and L. L. Sloss published the first geologically oriented computer program in a recognized geologic journal. Transistorized second-generation computers introduced. The ALGOL language was jointly introduced in several countries.
1961	Establishment of the symposia series "Computer Applications in the Mineral Industries" by University of Arizona.
1963	Announcement of third-generation microcircuit computers. First regular publication of geologic computer programs as Special Distribution Publications of the Kansas Geological Survey. First year more than 100 papers published on computer applications in geology.
1964	Time-sharing system successfully used at Dartmouth University.
1966	First series of geologic publications to deal exclusively with computer programs established by Kansas Geological Survey. First of eight colloquia on "Computer Applications in the Earth Sciences" sponsored by the Kansas Geological Survey. Establishment of an Associate Editor for Computer Applications for the AAPG Bulletin.
1967	American Association of Petroleum Geologists Committee on Electronic Data Storage and Retrieval formed. COGEODATA (IUGS Committee on Storage, Automatic processing, and retrieval of geologic data) formed.

TABLE 6.1 (Continued)

	Publication in IUGS Geological Newsletter, first international attempt to standardize description of mineral deposits in computer processable form.
1968	IAMG founded in Prague at the IGC.
1969	First issues of the Journal of the IAMG published. GEOCOM Bulletin, an international current awareness publication, initiated. US Geological Survey publishes its first Computer Contribution series. First book in a series on "Computer applications in the earth sciences" (published by Plenum Publ. Corp.)
1970	An informal research group on Computer Technology formed by SEPM.

6.5 PATTERNS AND TRENDS

For 150 years geologists have been collecting data. By its nature geology has been a historical and observational science. By the mid-20th century, however, this emphasis was beginning to change to one of understanding geologic processes (Sylvester-Bradley, 1972). This metamosphosis of where to how is being accelerated, and obviously the next logical step is one of understanding why, that is putting together the whole story. So, even in the short time geologists have utilized computers, a progression through several stages of computing environments has taken place (other disciplines have also undergone this transformation, for example, chemistry and physics).

Early applications were mainly analytical. Next, was a stage of collecting data in machinable form for use in predictive techniques utilizing methods developed and well tested in other disciplines as indicated in Table 6.2. Simulation followed and results are being evaluated in the 1970's. This progression in development of the subject is paralleled and recorded by examples in the literature (Preston, 1969). Past and future trends are shown in Figure 6.3.

TABLE 6.2 Historical Record of Stage of Integration of New Concepts or Techniques in a Discipline

	Publications	Data	Computer Programs	References
Discovery	Papers general with suggestions of possibilities	None	None	Practically none
Development	Papers demonstrate use of different techniques	Artificial	"Borrowed" from other fields intact	Mostly from other disciplines
Application	Papers acknowledge use of computers and source of programs. Different problems tried	Sample data sets	Modified and adapted from other fields with some geologic bent	Everything written on the subject in geology
Assimilation	Completely integrated	Real data in quantities necessary to solve problems	Programs written with only parts of "canned" programs used but specific for purpose	Citation of only those papers of pertinence to work

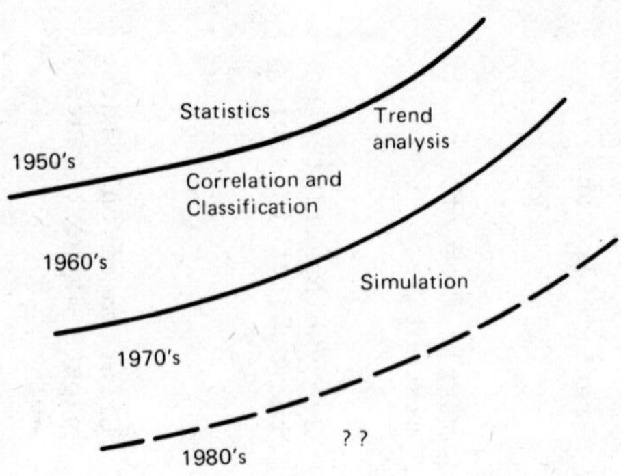

FIGURE 6.3

The future is naturally unknown. Developments are too rapid and users are heavily dependent on developments in the computer industry and other disciplines for advancement of methods and ideas. It is clear at this moment in time that we are moving from the application to assimilation stage in most areas and that we are using simulation to test real situations and learn of processes in an effort to understand why.

Past developments will have a bearing on future events. Individuals especially in universities have been developing and adapting techniques for solving geologic problems (mainly because of lack of funds and accessibility to data files). Simultaneously in industry, large-data files have been converted to machinable form. The development of standards and requirements for programs and data-file format by government agencies and other interested organizations have made considerable progress (e.g., Hubaux, 1970; Robinson, 1970). The wedding of new imaginative techniques as applied to real data of known quality in readily accessible, compatible files will result surely in a completely integrated system leading to significant findings. Only one warning. Geologists must keep in mind their objectives and define their problems sharply to keep from becoming a tool of the computer rather than the master.

REFERENCES

Harbaugh, J. W., and Merriam, D. F., 1968, Computer applications in stratigraphic analysis: New York, John Wiley & Sons, 282 p.

Hubaux, A., 1970, Description of geological objects: Jour. Math. Geology, v. 2, no. 1, p. 89-95.

Krumbein, W. C., and Sloss, L. L., 1958, High-speed digital computers in stratigraphic and facies analysis: Am. Assoc. Petroleum Geologists Bull., v. 42, no. 11, p. 2650-2669.

Merriam, D. F., 1965, Geology and the computer: New Scientist, v. 26, no. 444, p. 513-516.

Merriam, D. F., 1969, Computer utilization by geologists, in Symposium on computer applications in petroleum exploration: Kansas Geol. Survey Computer Contr. 40, p. 1-4.

Parker, M. A., 1952, Punched-card techniques speed map making (abs.): Geol. Soc. Amer. Bull., v. 63, no. 12, pt. 2, p. 1288.

Preston, F. W., 1969, Systems analysis--the next phase for computer development in petroleum engineering in Computer applications in the earth sciences: New York, Plenum Press, p. 177-189.

Robinson, S. C., 1970, A review of data processing in the earth sciences in Canada: Jour. Math. Geology, v. 2, no. 4, p. 377-397.

Sylvester-Bradley, P. C., 1972, Geobiology and the future of palaeontology: Jour. Geol. Soc., v. 128, pt. 2, p. 109-117.

Chapter 7

Problem Set in Geostatistics

R. B. McCammon

7.1 A PALEONTOLOGIST'S DILEMMA

Six fossils, Acanthus minerva, Gyro pipus, Rega elegans, Acanthus exerta, Rega veliforma, and Gyro robusta were found together in a tray at the museum. From an outside label, it is known that Rega veliforma, Acanthus exerta, and Gyro robusta are marine, fresh water, and brackish water species (not necessarily in that order) each from a different locality. Acanthus minerva was collected in Illinois. The fresh water species was collected in New England. Rega elegans is known not to be marine. The fresh water species, and one of the other, which is a marine species, were collected from the same locality. Gyro robusta is larger than the marine species. One of the species of the same genus as the fresh water species was collected in California. Which one is the marine species?

No doubt you will unravel the logic for this dilemma. Do you think, however, that you could write a computer program to solve the problem?

7.2 PARTICLE SIZE DISTRIBUTION IN THIN SECTION

In a recent paper by Rose (1968), it was shown that the probability p that an intersection figure cut at random in a block of thickness t from a single sphere of diameter D will have a diameter falling between the limits d_a and d_b is given by

$$p = \frac{D}{t}\left\{\sqrt{1 - \left(\frac{d_a}{D}\right)^2} - \sqrt{1 - \left(\frac{d_b}{D}\right)^2}\right\} \tag{7.2.1}$$

where $0 \leq d_a \leq d_b \leq D \leq t$.

1. Let D = t. Show that equation (7.2.1) is a probability distribution. What is the expected value? Prepare a frequency plot of this distribution.
2. Rose went on to show that if the number of spheres of diameter D equal to id was denoted as N_i where $d_a = (j - 1)d$ and $d_b = jd$, then the expected number between d_a and d_b was given by

   $C_{ij} = N_i \phi_{ij}$

   where

$$\phi_{ij} = \begin{cases} \frac{i}{m}\left[\sqrt{1 - \left(\frac{j-1}{i}\right)^2} - \sqrt{1 - \left(\frac{j}{i}\right)^2}\right] & i \geq j \\ 0 & i < j \end{cases}$$

   Write a computer program to generate the elements of the matrix ϕ for m = 10.
3. Suppose you are given the following observed frequencies of particle diameters determined from thin section:

Size class	Frequency
(smallest) 1	0
2	16
3	87
4	155
5	150
6	65
7	32
8	8
9	4
(largest) 10	1

Using the matrix generated above, determine the true frequency distribution of particle diameters. Interpret your results.

7.3 PARTICLE DIAMETER SAMPLING EXPERIMENT

We have just examined the probability distribution for the apparent diameter of a spherical particle of diameter D cut at random in thin section. Now let us simulate the process. Imagine a spherical grain

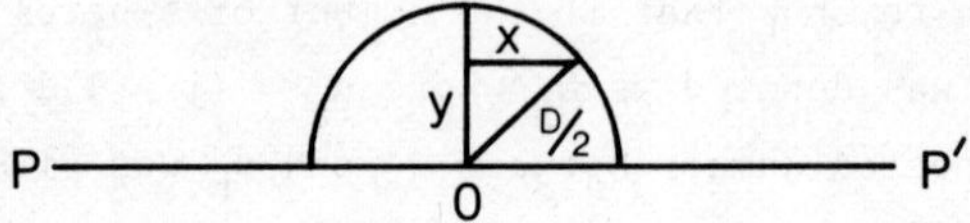

cut by a random plane parallel to PP'. Y is defined as a random variable distributed uniformly between zero and D/2, the particle radius. We have

$$p(y) = \begin{cases} 2/D & 0 \leq y \leq D/2 \\ 0 & \text{otherwise} \end{cases}$$

as the probability density of Y. The apparent diameter 2X related to Y by

$$X^2 + Y^2 = (D/2)^2$$

or

$$X = \sqrt{-Y^2 + (D/2)^2}$$

is a random variable also. We wish to find the probability density of 2X by experiment.

1. Let D/2 = 1. Use a random number generator function to generate a sample of 100 values of Y distributed uniformly between 0 and 1. Calculate 2X for each Y and prepare a histogram for 2X in intervals of 0.2. Calculate the mean and standard deviation. Retain the mean value.
2. Repeat the sampling process above 100 times, recording only the mean value of 2X. Prepare a histogram of the mean. How would you describe the shape of this distribution?

For a fuller explanation of the methods required to solve the two following problems, the reader is referred to McCammon (1969). The reference is given in Section 2.10 in Chapter 2.

7.4 LINEAR REGRESSION OF POROSITY DATA. I

The data given in Table 7.1 were gathered from well logs and rock cores taken from drill holes that penetrated subsurface formations in the Chicagoland area. At issue is the relationship between log-derived and core-derived estimates of porosity. The values given are expressed as percentages.

TABLE 7.1

Sample no.	Log-derived porosity	Core-derived porosity	Sample no.	Log-derived porosity	Core-derived porosity
1	10.0	5.5	26	10.0	9.6
2	9.0	3.6	27	5.0	10.3
3	7.0	3.6	28	7.0	4.5
4	6.0	4.9	28	8.0	6.0
5	9.0	7.1	30	9.0	6.7
6	7.0	2.0	31	5.0	4.1
7	10.0	8.5	32	8.0	4.5
8	5.0	5.2	33	9.0	6.5
9	7.0	2.6	34	7.0	3.3
10	0.0	1.9	35	3.0	2.1
11	5.0	6.1	36	7.0	2.5
12	6.0	9.3	37	7.0	6.8
13	9.0	6.9	38	10.0	3.7
14	8.0	4.3	39	7.0	6.0
15	5.0	3.3	40	5.0	3.4
16	6.0	2.5	41	8.0	2.2
17	6.0	4.8	42	4.0	1.8
18	5.0	2.4	43	5.0	2.9
19	8.0	3.8	44	8.0	2.6
20	15.0	18.4	45	16.0	15.3
21	16.0	14.7	46	4.0	16.9
22	7.0	10.9	47	5.0	15.7
23	12.0	12.5	48	14.0	12.4
24	14.0	18.6	49	21.0	22.9
25	22.0	22.1	50	21.0	21.8

1. As a first step, prepare a scatter diagram of the log-derived porosity versus the core-derived porosity values on graph paper by plotting the log-derived porosity value along the ordinate and the core-derived porosity value along the abscissa for each sample. Draw in the "best" fitting line by eye. Should the line be made to pass through the origin?
2. As measurements, core-derived porosity estimates are conceded to yield greater accuracies than log-derived estimates. However, rock cores must be analyzed for porosity singly using laboratory apparatus, whereas the acoustic log (the geophysical log commonly used for porosity determination) is recorded continuously with depth and therefore a continuous porosity profile in the borehole is generated. Now consider the log-derived porosity to be the dependent variable and calculate the intercept and slope parameters of the "best" fitting line using ordinary linear regression. Draw this line on the scatter diagram and compare it with the previous one.
3. It is known that the depth of investigation for the acoustic log extends beyond the borehole and into the formation. The transmitter-receiver distance for the tool is much greater also than the length of core used in porosity determinations. Thus, on the average, log-derived estimates of porosity may be as accurate (or representative) as core-derived estimates of porosity and thus both are dependent variables. Assuming the error estimates are approximately equal, calculate the intercept and slope parameters of the "best" fitting line. Draw in this line and compare it with the others.
4. Finally, past experience can guide us in establishing a fixed relationship between two variables subject to errors. Suppose it is known that core-derived porosity estimates are accurate to within one percent and that log-derived porosity estimates are accurate to within five percent. Calculate the intercept and slope parameters of the "best" fitting line and compare this line with the others.
5. There are obviously many approaches to analyzing these data. What is important to note is that prior knowledge after all is a part of the data.

7.5 LINEAR REGRESSION OF POROSITY DATA. II

On the assumption that the core-derived estimates of porosity are exact, we may wish to examine the nature of the regression, treating the log derived porosity as a dependent variable having independent and identically distributed normal random errors for successive observations. We can construct an analysis of variance table (Table 7.2) and proceed to test various hypotheses.

Let R^2 = (SS due to regression)/(total variation) be defined as the proportion of total variation explained by the regression. Calculate R^2 for both models. Perform an F test on the significance of the regression for each model at the 95 percent level.

For the linear model given by

$$Y = \alpha X + \beta$$

make a t test at the 95 percent significance level that $\alpha = 0$.

For the model given by

$$Y = \alpha X$$

make a test at the 95 percent significance level that $\alpha = 1$.

TABLE 7.2

Source of variation	SS, sum of squares	D.F., degrees of freedom	Mean square
Due to regression	$\Sigma(\hat{y} - \bar{y})^2$	1	
Residual	$\Sigma(y - \hat{y})^2$	n - 2	$s^2 = SS/(n - 2)$
Total variation	$\Sigma(y - \bar{y})^2$	n - 1	

7.6 SUNSPOTS AND EARTHQUAKES

It has been suggested (Simpson, 1967) that solar activity may be a triggering mechanism for earthquakes. Table 7.3, extracted from Fig. 2 of Simpson's article, covers the period between January 1, 1950 through June 30, 1963 and lists the average number of earthquakes occurring per day (>5.5 on the Richter scale) for those days in which the Zurich sunspot number fell within the stated intervals. A higher sunspot number is associated with greater solar activity.

TABLE 7.3

Zurich sunspot number	Average no. of earthquakes	Zurich sunspot number	Average no. of earthquakes
0-5	3.2	110-115	5.2
5-10	3.6	115-120	5.6
10-15	3.5	120-125	5.8
15-20	4.0	125-130	5.8
20-25	3.7	130-135	5.9
25-30	3.7	135-140	5.9
30-35	4.1	140-145	6.1
35-40	4.1	145-150	6.3
40-45	4.4	150-155	5.1
45-50	3.9	155-160	4.7
50-55	3.9	160-165	5.7
55-60	4.0	165-170	5.8
60-65	3.9	170-175	5.8
65-70	4.6	175-180	5.8
70-75	4.0	180-185	5.7
75-80	4.5	185-190	6.0
80-85	5.2	190-195	5.5
85-90	4.0	195-200	5.4
90-95	5.7	200-205	6.0
95-100	5.4	205-210	5.0
100-105	5.3	210-215	6.2
105-110	5.4	215-220	5.5

From these data, calculate the correlation coefficient between the daily average earthquakes recorded and the Zurich sunspot number. Prepare a scatter plot. Do you regard this correlation as significant? Does this imply a casual relationship? Is there an inherent fallacy in casting the data in this form?

7.7 A BEGINNING AND AN END

In a recent book, Shaw (1964) has proposed a new method for biostratigraphic correlation. For two fossiliferous stratigraphic columns, he has proposed that the time correlation be based on the first and last occurrences of the species present. Table 7.4, taken from Shaw, lists the elevations of first and last occurrences of fossil species for two stratigraphic sections from the Upper Cambrian Riley Formation, Llano Uplift, Texas.

TABLE 7.4

	Morgan Creek		White Creek	
Species	Base	Top	Base	Top
Kormagnostus simplex	299	485	460	655
Kinsabia varigata	373	464	561	653
Opisthotreta depressa	419	494	582	725
Spicule B	419	504	561	725
Tricrepicephalus coria	419	529	628	706
Meteoraspis metra	446	485	628	677
Kingstonia pontotocensis	453	475	628	706
Raaschella ornata	529	532	750	751
Aphelaspis walcotti	530	561	744	771
Angulotreta triangularis	538	570	756	779

Prepare a scatter plot of these data showing the elevation of occurrence of each species for the two stratigraphic sections. Calculate the correlation coefficient for:

1. First occurrences
2. Last occurrences
3. Combined occurrences

Do you think there are significant differences in the correlations? How would you account for these? How would _you_ go about establishing the equation of correlation for these two sections?

7.8 HELMERT TRANSFORMATION

The purpose of this exercise is to make you familiar with the geometrical interpretation of closed data arrays and to give you a feel for manipulating precentage data in three- or higher-dimensional space.

The Helmert transformation is defined for n-dimensional space by the matrix

$$P = \begin{bmatrix} 1/\sqrt{n} & 1/\sqrt{n} & 1/\sqrt{n} & \cdots & 1/\sqrt{n} \\ -1/\sqrt{2} & 1/\sqrt{2} & 0 & \cdots & 0 \\ -1/\sqrt{6} & -1/\sqrt{6} & 2/\sqrt{6} & \cdots & 0 \\ -1/\sqrt{n(n-1)} & -1/\sqrt{n(n-1)} & -1\sqrt{n(n-1)} & \cdots & (n-1)/\sqrt{n(n-1)} \end{bmatrix}$$

where the rows of P are the unit vectors of the transformation referred to as the initial Euclidean axes.

For an observation vector x $(x_1, x_2, \ldots, x_n)$ defined so that

$$\sum_{i=1}^{n} x_i = 1 \qquad x_i \geq 0$$

the origin first is relocated by defining

$$y_i = x_i - 1/n \qquad i = 1, \ldots, n$$

The transformed vector z $(z_1, z_2, \ldots, z_n)$ is defined by

$$z = Py$$

or expressed algebraically,

$$z_1 = 0$$

$$z_2 = 1/\sqrt{2}(y_2 - y_1)$$

$$z_3 = 1/\sqrt{6}(2y_3 - y_1 - y_2)$$

$$\cdots$$

$$z_n = 1/\sqrt{n(n - 1)}[(n - 1)y_n - (y_1 + \cdots + y_{n-1})]$$

Using triangular graph paper, superimpose the coordinate axes defined by the Helmert transformation for n = 3 on the triangular coordinate system. Plot the following points and check to see that the Helmert coordinates give rise to the same location as would be obtained using triangular coordinates.

	x_1	x_2	x_3
1	0.40	0.30	0.30
2	0.60	0.10	0.30
3	0.10	0.40	0.50

For the fourfold system, n = 4, determine the Helmert coordinates for the following three points in 4-space and draw the intersection figure for the tetrahedral representation of these points.

	x_1	x_2	x_3	x_4
1	0.40	0.00	0.00	0.60
2	0.30	0.00	0.70	0.00
3	0.20	0.80	0.00	0.00

Position the Helmert coordinate axes on the tetrahedron. What are the coordinates of the normal vector to the intersection figure formed by the three points?

For the following two problems, the reader should refer to McCammon (1969) for a more complete discussion of the methods involved. This reference can be found in Section 2.10 in Chapter 2.

7.9 A SPINE TO REMEMBER

The Florida crown conch (Melongena corona) is found commonly in the intertidal areas of Florida and Alabama, usually in the shade of mangrove trees. The shell is characterized by one or more rows of open spines on the shoulder. Colonial differences exist, however, which has given way to some preferring to erect subspecies. Imagine there are 46 specimens of the Florida crown conch in trays before you. Apart from the variation in size of the shell (as measured by its length), the most noticeable morphologic difference between specimens is the presence or absence of the development of lower spines. To aid you in the subsequent analysis, Table 7.5 lists for each specimen the shell length (expressed in centimeters) and the number (if any) of lower spines. Imagine that four of the specimens have been taped in such a way that the presence or absence of lower spines cannot be ascertained. These four specimens constitute unknowns which can be used to test the efficiency of any subsequent classification.

1. Prepare separate histograms based on shell length for shells which have and do not have lower spines.
2. There is clear evidence that the larger specimens tend to have lower spines, whereas the smaller specimens do not. The question is whether this difference has statistical significance. For the purposes here, we assume that the specimens represent a random sample from whatever parent population or populations we wish to define. We can perform first a t test of significance between the mean shell length for the two groups. Do this assuming first equal variance for the two populations and second, unequal variance.
3. Since a significant difference does exist, we can devise a discriminant function based on shell length that will predict the presence or absence of lower spines for a particular specimen. For these data, construct such a function and estimate the probability of a wrong prediction.
4. For the four taped specimens, predict for each whether lower spines are present.

TABLE 7.5

Sample no.	Length, cm	Number of spines	Sample no.	Length, cm	Number of spines
1	4.50	2	24	3.78	3
2	3.23	0	25	2.92	0
3	4.21	6	26	4.44	7
4	3.39	0	27	3.82	4
5	3.88	0	28	4.45	8
6	4.73	5	29	4.57	5
7	2.65	0	30	3.54	0
8	3.94	0	31	3.22	0
9	3.84	0	32	2.66	0
10	4.02	0	33	4.24	0
11	4.08	4	34	2.94	0
12	4.15	10	35	3.36	0
13	3.36	0	36	4.16	8
14	3.48	0	37	4.02	0
15	3.38	0	38	4.73	3
16	3.40	0	39	4.61	2
17	3.54	7	40	4.79	5
18	2.85	0	41	2.74	0
19	2.71	0	42	4.64	0
20	3.32	0	43	3.78	
21	4.59	0	44	4.09	
22	4.19	4	45	4.91	
23	3.44	0	46	3.96	

5. Having found the truth about the four taped specimens, (results on specimens available on request) add these to the sample data and construct a new discriminant function and estimate the probability of a wrong prediction.

7.10 NEARSHORE-OFFSHORE SEDIMENTS

The data in Table 7.6 represent grain size analyses for 17 samples of recent sediments collected from two separate environments and four additional samples that have not been classified.

1. Plot these data on a triangular diagram and draw in by eye the line which "best" separates the two types of environments.
2. Calculate the linear discriminant and draw in the line that represents this function on the triangular diagram. What is the probability of misclassification?
3. Classify the four sample unknowns as being either nearshore or offshore.

TABLE 7.6

Sample	Sand	Silt	Clay
	Nearshore sediments		
1	45	53	2
2	92	8	0
3	69	25	6
4	75	25	0
5	63	37	0
6	42	54	4
7	46	51	3
	Offshore sediments		
8	36	60	4
9	34	61	5
10	6	87	7
11	3	91	6
12	8	87	5
13	33	63	4
14	59	36	5
15	20	78	2
16	48	52	0
17	2	80	18
	Sample unknown		
A	19	71	10
B	64	35	1
C	33	55	12
D	21	74	5

4. Suppose it is known that the four unknowns were collected from the same locality. How would you now classify the group of four unknowns?

REFERENCES

Rose, H. E., 1968, The determination of the grainsize distribution of a spherical granular material embedded in a matrix: Sedimentology, v. 10, p. 293-309.

Shaw, A., 1964, Time in stratigraphy: New York, McGraw-Hill Book Co., 365 p.

Simpson, J., 1967, Earth and Planetary Sci. Letters, v. 3, p. 417-425.

Index